Woodworking

For Fun & Profit™

Jeff Greef

PRIMA HOME
An Imprint of Prima Publishing

PRIMA PUBLISHING and colophon are registered trademarks of Prima Communications, Inc.

The FOR FUN AND PROFIT logo is a trademark of Prima Communications, Inc.

Disclaimer
Some of the projects discussed herein involve the use of power and other potentially dangerous tools. The improper or unsafe use of such tools can result in serious bodily injury or death. These tools should be kept away from and out of reach of children. Also, it is important to follow the manufacturer's safety recommendations and use standard safety precautions, such as the use of proper safety glasses, in using such tools and creating the projects described in this book.

Prima Publishing and the author hope that you enjoy the projects discussed in this book. While we believe the projects to be safe and fun if proper safety precautions are followed, all such projects are done at the reader's sole risk. Prima Publishing and the author cannot accept any responsibility or liability for any damages, loss, or injury arising from use, misuse, or misconception of the information provided in this book.

Library of Congress Cataloging-in-Publication Data

Greef, Jeff.
    Woodworking for fun & profit / Jeff Greef.
        p.   cm.
    Includes index.
    ISBN 0-7615-2038-4
    1. Woodwork. I. Title. II. Title: Woodworking for fun and profit.
TT180.G64 1999
684'.08—dc21   99-39231
                                                                                CIP

99 00 01 02 ii 10 9 8 7 6 5 4 3 2 1
Printed in the United States of America

How to Order

Single copies may be ordered from Prima Publishing,
P.O. Box 1260BK, Rocklin, CA 95677; telephone (916) 632-4400.
Quantity discounts are also available. On your letterhead, include
information concerning the intended use of the books and
the number of books you wish to purchase.

Visit us online at www.primalifestyles.com

Contents

The FOR FUN & PROFIT™ Series

Decorative Painting For Fun & Profit

Holiday Decorations For Fun & Profit

Knitting For Fun & Profit

Quilting For Fun & Profit

Soapmaking For Fun & Profit

Woodworking For Fun & Profit

Introduction

WOODWORKING IS ONE OF our most popular hobbies, and for good reason. With only a minimum number of tools and a little self-training, you can make beautiful, useful items that bring you a real sense of satisfaction. "Woodworking" is really an umbrella term for a wide variety of craft possibilities that will suit an equally wide variety of tastes, interests, and abilities. From making reproductions of fine antiques to simple whittled figurines, complicated machine techniques to basic hand-tool methods, freehand bowl turning to precise dovetail jigs, there is something for anyone with a love of wood and an urge to create.

Perhaps the greatest benefit you can derive from your woodworking craft is the satisfaction that comes from making something yourself. There is the enjoyment that comes with the process itself—using your tools and watching your piece take shape. You'll have a few frustrating experiences here and there, but solving these problems in itself gives a sense of accomplishment. Then you'll feel the great sense of pride that comes when you finish that first tough project—the one you weren't sure you could do—but there it is, showing off your newfound skills. When you show it to other people, will the smile be bigger on your face or on theirs?

There are many good reasons for you to take on or expand your woodworking hobby. Such a wide range of project possibilities exists that you can pick whatever skill level suits you best. You might decide to focus on one simple thing, like turning round candlesticks on a lathe or making small boxes, and never explore much else in woodworking. Or you might decide to learn a wide variety of skills and apply them to create elaborate reproductions of antique furniture or your own modern designs.

Woodworking doesn't have to involve a huge investment in tools and materials. Your degree of investment will be a function of what you want to make, or perhaps you will limit what you make to the tools and materials you have available. Some people do what they want in a corner of a garage, whereas others have dedicated shop spaces of varying sizes and descriptions. You can pick whatever suits you.

Another great benefit to choosing woodworking as a craft is the amount of information available for you to learn from. Numerous magazines are devoted to various aspects of woodworking, and many books are published on a wide range of topics. When I got started, I gobbled up every page of *Fine Woodworking* magazine when it came in the mail, as I was fascinated by the range of techniques I was learning about. Any beginner will do well to subscribe to a few magazines and tank up on all the good ideas found in them. Many woodworking videos are available, and Web sites on woodworking are out there as well.

I was somewhat daunted when I got started with woodworking about 15 years ago because the techniques and skills I saw in the magazines looked so difficult and complicated. How could I ever learn to do that? But I fell back on what my father, a woodworker like his father and grandfather before him, taught me in our garage when I was young: You take it one step at a time. The first time I tried making cabriole legs for a small coffee table, I had no idea what I was doing, except for a procedure I had read about in a magazine. The first step was to cut them out on the band saw. Okay, that wasn't so hard. Then I had to shape them on a stationary sander. That was tricky, but you get the knack by doing and being willing to risk making a mistake. I made a few ugly divots in the first leg with the sander, but I just smoothed them out again. Gradually, I ended up with four smooth and attractive legs on a nice little table. I donated it to a charity auction, and the buyer looked at me like I was a pro. Why tell him about my nervous anticipation before making the piece when, after all, it turned out just fine?

With the experience that I have now, I rarely feel the same kind of nervous anticipation that I felt when I was trying things for the first time. When I do a new procedure, it's a challenge: I enjoy applying my skills in a direction I haven't gone before. For me, exploring new design avenues and finding the best techniques to accomplish them is the most satisfying aspect of my craft. Seeing a unique object that I conceived of, developed techniques for, and then completed gives me a sense of individualism, which I think many people find hard to obtain in our modern, global society.

Woodworking Past and Present

FASHIONING WOOD INTO IMPLEMENTS that are useful in our daily lives has a history that goes back as far as the archaeological record. The stone, copper, and iron axes and adzes used by prehistoric peoples undoubtedly were utilized for much more than chopping firewood. Then, as now, wood was a readily available and easily worked material that was well suited for making houses, bowls, boxes, chairs, ships, weapons, and innumerable other things. A high level of trade, based on wooden boats, existed throughout the Mediterranean area 3,000 years ago. Examples of fine furniture taken from tombs in Egypt and elsewhere show that the level of skill attained in early times was very high. You might be surprised to learn that the Egyptians used plywood, laminating together pieces of thin veneer for various purposes.

From the Middle Ages until the industrial revolution, we see a gradual increase in complexity and stylistic diversity in European woodworking. Examples from the early Middle Ages show simple if rugged construction, occasionally ornamented with much carving. Although the carving was sometimes elaborate, the basic design was not, so that a useful item could be produced quickly—necessary in hard times. Through the Renaissance until the industrial revolution, we find more elaborate basic furniture designs that incorporated, for example, curved table and chair legs. It's true that the finest furniture styles could be afforded only by the nobility, and at times they certainly produced excessively ornamented pieces that were very costly. However, the middle class, which was on the rise as well, was able to afford less fancy furniture that was both pretty and functional. The term "cabinet" literally means "small cabin" or "little house with glass." As the middle classes acquired enough resources to collect things like fine china, they needed a way to store these items and show them off—thus the china cabinet.

Until the industrial revolution, all woodworking was done by hand; woodworking was a skilled but very time-consuming profession. With the advent of machinery, the face of the crafts changed forever. It was then possible to mass produce furniture at such low cost that the old-style craftspersons simply couldn't compete. But mass production came at a cost. Much of what was produced was very shoddy, put out fast and in great quantity to supply the desires of a growing population. In addition, mass production doesn't lend itself to customized work or quirky, one-of-a-kind projects. Factory woodwork offers a limited range of designs to choose from.

People recognize this mass-produced look and seek out something different. That's where you, the independent woodworker, come in with a good product that is not made in large factories. However, this doesn't mean that you must use only old-style hand-tool methods, although this can be very satisfying. The industrial revolution has also brought the independent woodworker the benefit of a wide variety of small, inexpensive woodworking machines and other devices that can do a lot of work quickly.

Using This Book

IF YOU DON'T YET have a specific item in mind to make with your woodworking craft, use this book to get ideas for techniques and the types of woodworking you might want to consider. Artists often conceive of an idea for a project and then find a way to carry it through, but sometimes it happens the other way around. You might be looking at a certain technique that you weren't aware of, and the design possibilities that it presents give you an idea for something to make. Perhaps you already are an experienced woodworker but want to refresh or expand your horizons. This book might show you alternate ways of doing what you have been doing, or you might discover a new direction altogether.

Once you feel more comfortable in your craft, you might decide to make some money with it, along with gaining personal satisfaction! You'll find chapters that give you basic information on lumber, how-to information for a range of woodworking techniques, and specific projects that you can market. The second half of this book is devoted to getting you going on the business end of things, with chapters on general business operations and dealing with outlets (craft fairs and galleries) for selling your work. The book ends with a resource section that shows you where to find more specific information on topics that interest you.

Perhaps the most important thing to remember as you read this book is that you need to find your own woodworking niche. Find the kind and level of woodworking that suits you best. You might already have advantages in your favor, such as materials and tools on hand, that make it easy to go a certain direction. If you don't have those advantages, but if you know the direction you want to go, it will be worth your effort to get what you need. The point is to look at what you want to do, determine what you have on hand or need to acquire to do it, and then find the route that works best for you.

It might take some time, but you'll find your niche and soon be happily working away!

Part One

For Fun

The Joy of Woodworking

PROBABLY THE MAIN REASON that woodworking is so popular today is that it is so easy to become involved and obtain real rewards for your efforts. In a high-tech world in which so much of what we use is made by machines beyond our reach, it's nice to know that there's still something the individual can do by him- or herself that isn't designed and manufactured by nameless people miles away. You have no control over what comes out of the greater marketplace, Washington, or your TV, but you do have control over what comes out of your woodshop.

Many people feel that designing and creating their own crafts is an excellent means of self-expression. The pleasure you'll get from looking at a piece of woodworking that you made yourself is unique indeed, and the relaxation and simple enjoyment that come with woodworking are a great way to relieve the stresses of modern living.

To be a woodworker, you need to be a problem solver, and this is another of woodworking's rewards. Put ten woodworkers together to solve a woodworking problem, and you'll get fifty solutions. It's that good old sense of American ingenuity; we pride ourselves on our self-reliance and adaptability.

The Range of Possibilities Open to You

You'll manifest your own ingenuity and adaptability in the type of woodworking you choose to pursue. Your choice can be based on a number of reasons, from fulfilling a need that you see expressed by others to working on a project simply because it appeals to you. Your choice might be based on the available tools and materials, or you might seek out the tools and materials you need for a specific project. This chapter outlines the types of woodworking that you can pursue; all you need are the right instructions for learning and the proper tools. This outline is by no means complete, but it's a good start. Don't be afraid to try something new or unusual, as doing so often will bring you the most satisfaction and really catch people's eyes.

Did you know???

The earliest description of woodworking procedures comes to us from Homer, who, in *The Odyssey*, details how Odysseus built a small boat by planing boards smooth, laying them out like a shipbuilder, and boring holes into them for pegs.

Boxes

Making boxes of various kinds has been popular with woodworkers for a long time. Everyone likes them, and we use them for so many things. They are perfect for the hobbyist woodworker because making them requires only a small shop and very few tools. The design possibilities are almost limitless.

Box making can take a fair amount of time, depending on how detailed your joinery is, but designs for small boxes often lend themselves to production methods that can reduce the time required to build them. So, if you are planning to make one nice box for Aunt Mathilda, you can make five more at the same time for next year's Christmas presents. The result will be that you spend less time per box than you would making only one.

Lathe Turning

A lathe is an unusual woodworking tool, as it does something that no other tool does: It makes a piece of wood spin around, allowing you to apply a tool to the wood to cut its surface and change its shape. Lathe turning offers a great outlet for creativity and personal expression. You don't need many tools—just a lathe and perhaps a band saw or table saw to prepare your work for the lathe. The lathe accommodates a variety of attachments and allows you to produce finished pieces in less time than do many other kinds of woodworking. The piece simply goes onto the lathe and is worked and finished on the tool until it is completed.

> ### Did you know???
> Ornamental lathes were developed over 200 years ago as a pastime for rich nobility. At the time, lathes were very expensive, costing as much as a block of row houses in London. One of the old Holzapfel lathes now costs from $10,000 to $20,000 or more.

Lathe turning is similar to sculpting, the goal of which is to make the most pleasing shape on your curved surfaces. It also involves unique skills that don't apply to other types of woodworking. You hold a tool against a spinning piece of wood and freehand the cut as you go. The design possibilities are limitless, and each turner must find his or her own approach and style.

A wide variety of items can be made on a lathe, including bowls, plates, and cups. Legs for tables and chairs can be lathe turned in round, symmetrical shapes or by offsetting the turned piece in the lathe to achieve an asymmetrical form. Candlesticks are quickly and easily turned out, and some people have made their livings simply from mass producing these.

In art and craft magazines, you can find a variety of lathe-turned objects that are a kind of sculpture. Some artists begin with a large piece of wood that has been turned and then work the piece further by carving it or adding other things to it. Some artists begin by laminating together large chunks of various woods of different colors and shapes and then turning that piece. The result is similar to an inlay with many small pieces of wood carefully fitted together. Sculptural turning makes good use of unusual pieces of wood, such as burls or other gnarled chunks.

A specialized area of lathe work is called ornamental turning. A piece of wood is placed in a small lathe and cut until it is cylindrical, but then it's held still in the lathe while special milling attachments with small spinning cutters are used to cut small facets into the piece in geometrically complex shapes.

Woodcarving

Woodcarving offers a wide variety of possibilities. You can carve sculptures in any size and shape. One man laminated together enough wood to carve a life-size head of a bull elephant, and in another project he carved about half a full-size biplane with the aviator standing alongside it. Such sculptural carving, large or small, can be in the form of panels that fit in a door or that grace a living-room wall. Carving has long been used to embellish furniture and other kinds of woodworking to give attractive highlights.

Running carving is a technique by which the carver achieves a repeated geometrical pattern along a given surface. You make the same cut over and over as you "run" along the length of the piece. Then you start over at the beginning with a joining cut (or cuts) that results in the pattern you want. This can be used to dress up the edge of a tabletop or to make large perforated screens for room dividers.

Children's Furniture

Here's an idea that's always popular. Little folks need little chairs and tables, and dressing them up in fun and attractive designs gives you a chance to be creative and whimsical. Of course, you can scale down other basic furniture designs to the proportions of children, but why not add some flair and individuality by designing a chair out of plywood that looks like a cat or dog or making a bedstead with a big animal's face on it?

Toys and Puzzles

I once made a wooden bulldozer and dump truck for my nephew, though I thought these toys would have little chance competing for his attention with all his fancy plastic toys. To my surprise, his mother told me that these wooden toys were his favorites that Christmas. He had become so bored with the flimsy, multicolored plastic toys that he saw the solid, wooden ones as novelties.

Many books are available on making toys and puzzles of all kinds. With a scroll saw (sometimes called a jigsaw), you can cut out jigsaw puzzles in any shape you want. You can also make building blocks in a wide variety of shapes and sizes, giving children their own creative freedom to build things.

Picture Frames

Local picture frame shops offer a limited number of frame types and styles, but you can make frames that are more attractive than the simple metal and wood frames found in those shops. You can use carving to embellish frames or hand pick very attractive pieces of wood to give a unique look.

Household Items

You can make cutting boards, napkin holders, salt and pepper shakers, spice racks, and many other small items for the house. Of

Handy Hint

With children's furniture, remember that you need to consider some safety issues. Round off all corners well so that if a child trips and falls on it, he or she won't be injured by a sharp corner. If the children using the furniture are so young that they might put their mouths on it (as very young ones do), use a safe finish, such as the walnut oil finishes sold for use on salad bowls, or shellac, which is edible when dry.

course, all these things are available inexpensively in stores, but you can make these items more attractive than store-bought ones. By using a special kind of local wood that is beautiful and can't be found anywhere else, you can create something truly unique.

Furniture

How many times have you looked at a piece of furniture and said, "Well that's nice, but I'd prefer that it didn't have this and did have that." You can take those preferences and turn them into your own furniture style. Creating your own style gives you the opportunity to freely express your design ideas. You can develop these ideas by looking at designs from the past and changing them or adding to them to suit yourself, or you can start from the ground up and re-think the whole idea of furniture.

However, remember that rethinking the whole idea of furniture without regard for the lessons of the past can be filled with pitfalls. Designers of yesteryear had good reasons for designing as they did. Furniture design is always a combination of form and function, looking for a way to make a structurally sound, functional item that looks good as well. Thus, designers have always sought ways to adapt basic structural techniques to the design direction they wanted to follow. Many designers are adapting styles or elements of styles from the past into their work because these are familiar and pleasing. The best contemporary furniture makes old ideas new again by incorporating them into a modern idea.

Rustic

This style incorporates rough sections of wood into the finished piece, such as tree branches, log sections, and so on. Here's where interesting pieces of lumber that you find, such as twisted driftwood or whatever you find on the forest floor, can be put to good use.

You'll need to look long and hard at the joinery you use for this kind of furniture, as it can be tough to join two pieces of wood that aren't straight or smooth. But special tooling, such as tenon cutters that cut the end of a tree branch into a uniform dowel shape, can help you accomplish this.

Bentwood or Wicker

This, you can say, is a type of rustic furniture. The idea is to take pliable branches, say, of willow and bend a large number of them around to the shape of a chair or table and then carefully nail them together. Obviously, you'll need a ready source of green willow branches to do this, but you don't need much else.

Country

Here's a simple style of furniture that probably has its roots in colonial furniture. The emphasis here is on basic designs and on rough or knotted wood that gives a rustic feel. Straight, rectangular designs mean practical, economical construction. Chairs and tables in this style often use lathe-turned legs, a quick way to make an attractive piece.

Did you know???

When Columbus came to America, the eastern forests of the North American continent extended from Maine to Florida, and inland all the way to the Mississippi. It is said that a squirrel could have traveled this entire expanse without touching the ground.

Deck and Lawn

If you look at the offerings in local stores for deck and lawn furniture, you find either really cheap stuff that looks, well, cheap, or you find nice things that are very expensive. Your outdoor furniture is going to look at least as good as the better factory-built items and will cost you a lot less. Stick with simple designs for furniture that will be exposed to the elements, as fine details and small joints won't hold up to the weather as well as broad joints held with numerous waterproof screws.

Shaker

The Shakers were a New England religious sect that emphasized simplicity in their lives and works. They were excellent craftspeople who produced furniture in their own style, related to colonial and country styles in their simple, basic designs. Shaker-inspired furniture can be efficient to produce because of its simple designs, and its popularity has spread in recent years.

Arts and Crafts (Mission)

Toward the end of the Victorian period, a sort of rebellion took place among many craftspeople against shoddy, mass-produced furniture on the one hand and expensively ornate styling on the other. The arts and crafts movement responded to these sentiments with furniture that was simply but sturdily designed so that it would be economical but not shoddy, attractive but not gaudy. The mission chair is one of the most recognizable manifestations of this style. Using many straight lines and basic designs, this style of furniture is not very difficult to make and will probably always be in style.

Reproduction Antiques

Not all furniture from the past was extremely fancy and difficult to build. The basic Windsor chair is a fairly simple piece to make, although developing the skills to do so takes some practice. But much of the construction is done with simple hand tools and by eye alignment, so making them is not as much a matter of fine precision as it is of practical accuracy. That's because the original makers were trying to make an economical chair, and so are you. Colonial furniture was simple and economical because that's what people needed at the time.

However, some woodworkers go the long route and produce beautiful reproductions of difficult pieces, such as Queen Anne chairs and tables with hand-shaped legs or secretary desks such as the ones America's founding fathers used. These projects are time consuming and require much skill, but they can be done in a small shop just as well as other woodworking can.

Commercial Work

Finally, let's take a brief look at a couple of types of woodworking that you are more likely to do for hire than just for yourself. When looking at doing custom cabinetry, doors, and windows, you need to think about the business end of things from the start, as you will be competing with established shops.

Cabinetry

Probably most small woodshops across the country are devoted to producing mainly custom cabinetry. This is because factory cabinets come only in certain sizes and thus don't fit well into everyone's kitchen and bathroom. Cabinets are the first thing you see in a house and what you use most frequently, so people want attractive cabinets that don't look cheap.

But it's also true that custom cabinets are very time consuming to build and thus expensive. Consequently, the business is highly competitive, as many customers shop mainly for price. Many small shops are set up to knock out cabinets in a production fashion using special tools and fasteners. You can't compete with these shops because they have the facilities and methods that will make them faster and more efficient than a small, part-time shop.

Woodworking Magazines

Here's a brief look at some of the major wood magazines and what they offer.

- *Fine Woodworking.* The oldest and most respected magazine devoted only to woodworking; a good place to learn basic and more advanced methods; emphasizes project techniques.

- *American Woodworker.* Project articles and techniques for the beginning, intermediate, and advanced woodworker.

- *Woodwork.* Project articles, techniques, and interviews for all woodworkers.

- *Woodshop News.* Monthly tabloid with hard news for commercial woodshops; includes interviews and information on tools and techniques of interest to all woodworkers.

- *Woodworker's Journal.* Project articles and techniques for the hobbyist.

- *Workbench.* For the home do-it-yourselfer, a broad range of topics, from hobbyist projects to home repair.

- *Woodsmith.* Project articles and techniques for the hobbyist.

- *Today's Homeowner.* Home repair topics, from building a deck to fixing electrical to any topic on owning a house.

- *Popular Mechanics.* Devoted mainly to nonwoodworking subjects but always includes woodworking articles on projects, tools, and techniques.

Still, a definite niche exists for the small custom woodworker in building custom cabinets. That niche is usually doing smaller, higher-quality built-ins. The competitive shops make working cabinets, but you get what you pay for, and often customers look for something better and are willing to pay for it. That's great for you because it's more fun to make better-quality cabinets! This niche can include making one or two cabinets to match existing ones or some special design for a unique purpose that doesn't fit the program for the local production shops. To do this kind of work, you must be very adaptable to working in different designs and with different techniques. Getting this kind of work is usually word of mouth; once you do a few jobs, the word gets around.

Entry Doors and Custom Windows

In a small shop working part time, you can't make a houseful of custom doors and windows, but you can make an entry door or one or two custom windows. People are willing to spend a bit extra on the front door of a house to make it look nice, and all of us would like to have a pretty stained-glass window in our living room. The big sash and door factories have very limited offerings, and even local woodshops that make doors and windows are often reluctant to take on very small jobs involving only one or two doors or windows because they need to do larger production runs to pay their larger overhead costs. You, however, have much lower overhead and can take these small jobs, and you can work on custom designs that are fun for you but not always practical for large shops.

The techniques and vocabulary of custom wood sash and door building are specialized, but you don't need large shapers and expensive shaper cutters to make the required stile and rail joints. This can be done with special router bits on a router table. Getting this kind of work can be word of mouth, but you can contact local stained-glass artists who, from time to time, will do jobs that need new windows or doors and their frames (called jambs). You might leave your name with local sash and door shops, as they might want to give your name to customers whose jobs are too small for them. Professionals might laugh at "someone working in a garage," but the fact is that the smallest shops can give the most personal attention to a situation and fill the niche that requires this attention.

Conclusion

The recurrent theme as we look at all these different types of wood-working is that you must apply your creativity to whatever you are doing so that you can make what you are doing unique and personal. Some kinds of woodworking have the goal of reproducing what was done in the past, and with this kind of work your personal expression is found in how well you execute an established form. That is a valid and fun kind of work, but at the same time many of us got into woodworking because we wanted to express our own ideas. For that you need the spit, spunk, and daring to try something new and see how it flies. That's the joy of woodworking!

chapter

2

Getting Started

THIS CHAPTER CONTAINS INFORMATION on the materials and tools used in woodworking. What you'll need for any particular project will be dictated by the type of project you decide to do, and you'll likely need materials and tools that are not mentioned here. Nevertheless, this chapter will serve as a good place to start, and it might give you some ideas about ways to find wood that you hadn't thought of before or tools and techniques that you haven't tried.

Lumber

Lumber can be obtained from a variety of different sources, and it comes in many forms. Although lumber can be bought from many places other than the local lumberyard, for many people this is the only ready source. Let's take a look at what you will find at a lumberyard.

Hardwoods

Most hardwood lumber comes in two forms: surfaced and unsurfaced. Surfaced lumber comes off the sawmill at 1 or 2 inches thick and is then surfaced (planed) to about ¾ or 1¾ inches thick, respectively. This surfacing removes the saw marks from the mill. That's why 1x (one-by) lumber is really only ¾ inch thick and 2x is less than 2 inches. You can also buy unsurfaced lumber (full-sawn lumber), which is the lumber as it comes off the sawmill. Sometimes, this is less expensive, but you need a way to surface it. People often buy and use surfaced hardwoods at the thickness they come because it's easier to work with.

Hardwoods usually come in random widths and lengths because hardwood trees are generally small and cutting them up yields pieces of different sizes. Rather than make all the pieces the same size and then waste the extra, lumberyards sell it to you as big as they can and let you cut it up. Generally, though, they straighten one edge and leave the other rough or curved.

Hardwoods are usually sold by the board foot. A board foot is a volume of wood that is equal to 1 foot square and 1 inch thick. Surfaced wood is sold at the full thickness that it started at (i.e., 1 or 2 inches), not the actual thickness you get. It's similar to cereal settling in the box—you pay for a big box, but lo and behold, it ain't all there!

Softwoods

Softwoods usually come in nominal sizes: 1 or 2 inches thick by 4, 6, 8, 10, or 12 inches wide. This is because softwood trees tend to be large and the mills can cut out whatever size they want. It's possible to get softwoods in unsurfaced form, but usually they come surfaced. One-by softwoods are ¾ inch thick, and 2x is 1¾ or 1½ inches thick. Nominal softwoods are priced and sold by the lineal foot.

Lumber Grades

Lumber is graded by its cut and quality. Cut refers to how it was cut out of the tree, either plain sawn (also called flat grained) or quarter sawn (also called vertical grained). Both refer to how the annual rings are oriented to the width of the board. Flat-grained sticks have the rings parallel to the wide face, and vertical sticks have the rings at 90 degrees to the wide face. When they are at about 45 degrees, it's called rift grained.

It's more difficult for the mills to cut out lumber that's vertical grained, and they get less lumber from the tree, so it costs more. However, it's desirable for them to cut vertical grained lumber because it's more stable (see the section "Wood Movement") and has a different look. But on some woods, flat-grained lumber has the appearance you want.

The quality of the lumber refers to how many knots, splits, or other defects are in each board. The more defects, the less useful the board. The best-quality hardwood boards are called "firsts and seconds" or "select"; various lesser grades are called "common."

Kiln Drying

All lumber in a lumberyard has been placed in a kiln and dried to a certain extent. Construction lumber (which is always too nasty for anything except building walls in construction) is dried minimally; that is, most of the moisture in it is removed, but it's still somewhat wet. Hardwoods and better softwoods have been dried further, to about 8 percent moisture content. You don't need or want the wood dried beyond that because the wood will soak up moisture out of the air.

Exotic Species

Some lumberyards carry unusual and beautiful species of wood, such as mahogany, teak, bubinga, and purpleheart. Unfortunately, many of these woods are being harvested in unsustainable ways in the forests they come from, which simply means that the

> ## Handy Hint
>
> If you need a lot of short pieces of lumber, buying lesser grades can save you money because you can cut around any defects. However, you'll end up throwing away a lot of small pieces, so before you buy knotted lumber, be sure that you can use it efficiently.

loggers aren't taking care of the forest so that it can rejuvenate. Fortunately, a worldwide movement to promote the sustainable harvesting of lumber in all forests, including those in the United States, is taking place. Credible organizations will certify that lumber taken from a given forest has been harvested sustainably so that you know that the lumber you are buying is "good wood." The certification process is beyond its infancy, and it is well established in many places but still new in others. Ask the lumberyard personnel about certified lumber and tell them that's what you prefer.

Milling Local Trees

You might have good lumber in your backyard. Some beautiful local species might be unavailable in a lumberyard. A local sawmill could cut a tree up for you if you pay to have it hauled there. You could cut it up yourself with an Alaskan chainsaw mill, which is a large chainsaw that has attachments for slicing a log into slabs. However, this is time consuming, and unless you're very careful, the slabs might not come out straight. You could also have someone with a mobile mill come to where the tree is and cut it up. They will charge you a hefty fee, but in a day you'll get a lot of lumber if you have a big tree. The two types of mobile mills are the band-saw type and the circular-saw type. The band-saw type is best for hardwoods because it has a minimal kerf (thickness of cut), so it wastes less wood. The circular-saw type is best for cutting up softwood trees, as it will quickly produce a large quantity of lumber, but the large kerf wastes a lot of wood and makes a huge pile of sawdust.

If you do this, get a metal detector and pull out all the nails and iron stuck in the tree; otherwise, the operator will curse every time he or she hits one. Nails mess up the blades in a hurry. Stack the lumber out of the sunlight, and put ¾-inch-thick stickers between all layers of the stack so that air can circulate around the wood as it dries. It takes one year per inch of thickness for the wood to become air dry, which means that it has lost all the moisture it can, given how much moisture is in the air in the form of humidity. Air-dried lumber is okay to use for any purpose, unless you know that the finished product will be placed in a location of significantly lower humidity, in which case the wood likely will shrink a lot and cause problems. You can have green lumber dried in a kiln if someone in your area has a kiln and is willing to do it for you.

Don't put green lumber in the attic. It will dry too fast and then split because the outside surfaces will lose moisture far faster than the inside. You want to dry it slowly. Once the wood has been air dried, you can put it up in the attic for the summer; this will guarantee that you've safely driven out as much moisture as possible.

Before you have a tree cut up, be sure that it's a usable species. Some woods, such as eucalyptus, twist and split so much when they dry that they become useless. Others might be so plain looking that it's not worth the effort. Also, be sure that the log isn't rotted or split to begin with.

> ## Handy Hint
>
> Always put paint (any kind will do) on the ends of green lumber. This reduces splits in the ends of the wood as it dries; the splits result from the open pores on the end grain losing moisture faster than the rest of the wood.

Found Lumber

You can find all sorts of interesting pieces of wood at the lakeshore, seashore, or wherever trees grow. You'll find pieces in odd shapes and sizes that no lumberyard has for sale. Such wood can best be used only for special projects, such as turnings and rustic furniture, but let your imagination run wild!

Recycled Lumber

As the price of lumber increases, interest in salvaged lumber does as well. Some landfills are sorting out usable lumber from the waste brought in. Whenever possible, demolition companies save large beams and other flooring or framing lumber from old warehouses or other buildings. Some of this wood is beautiful old-growth lumber, and it can be worthwhile to have salvaged beams cut into smaller sizes with a band-saw mill to yield furniture-grade lumber. You'll likely find knots in this kind of lumber, and you should once again get out your metal detector and pull those nails.

Problems with Lumber

There are many things to look out for when you buy lumber. End splits occur in the drying process, and although the split might appear as only a crack a few inches long, it might continue much farther, making the board much less usable. Inspect boards for straightness. You can cut a warped or bowed board into smaller pieces to make them reasonably straight, but the only way to really straighten an unstraight piece is with a jointer and table saw or a planer. If you need long, straight pieces, it's best to start with long, straight pieces rather than twisted ones. Any lumber that you buy outside a lumberyard should be inspected for insects. I once had a bay tree cut up, only to find that there were beetles throughout the sapwood, which I had to cut up and throw away. If lumber has been stored outside without being stacked with stickers between each layer so that air can circulate, water damage in the form of fungi or rot can occur. The stickers allow air to evaporate moisture before fungi set in.

Watch out for being shortchanged as well. If you buy a large quantity of hardwood in random widths and lengths, the lumberyard might pull the lumber for you, tally the amount of wood, and ship it to you. Because the wood comes in random sizes and curvy edges, such tallying can be somewhat subjective. Your total tally of board feet might be very different from the yard's. Be aware, however, that they usually deduct about 10 percent for a straight-line fee, meaning that they charge you a bit for having cut a straight line onto one of the edges of each board. Still, do a careful tally and don't be afraid of browbeating them into giving you a partial credit.

Wood Movement

Wood movement is an issue that woodworkers must always be aware of. As wood takes on or loses moisture from the air, it expands or contracts, respectively. But unlike the sponge in your kitchen sink, wood does not expand and contract uniformly in all directions. Virtually no change takes place along the length of the piece, but the cross-grain dimensional changes can be enough to cause serious failures if you don't design your piece to allow this movement to occur safely. Also, most woods expand and contract more tangentially to the annual rings in the tree than they do across them.

Did you know???

A tree that leaned as it grew will yield unstable wood (reaction wood). This is because as the tree grew, it built tension into the wood to support itself. This tension will translate into warped and twisted boards.

Wood movement is a serious issue only when you have wide pieces of lumber, from 6 inches wide on up. If you edge glue five pieces of red oak together to make a tabletop 36 inches wide, that top can easily change in width by ¼ inch or more as the seasons change and humidity in the air goes up and down. If that top is glued or solidly screwed down, the expanding or contracting wood will break something, even itself, as the wood does what it must do. In this case, the solution is to attach the top with slotted keepers that can slide a bit as the top moves.

Another problem can occur with that red oak tabletop. Remember that wood moves more tangentially to the rings than across them. If those five boards are all flat grained, and if you glue up the top such that the tree centers for all the boards are located on the same side of the top, it will cup. This means that it will curve up on the ends or in the middle, becoming U-shaped across the width. If they all cup the same direction, disaster will result. If you orient the tree centers on adjacent boards on alternate sides of the top, the result of cupping will be a serpentine shape, and the cupping won't be as noticeable (see figure 1). Remember that you can also buy vertical-grained lumber. This grade will not cup like flat-grained lumber, and it moves less along its width than does flat grained. It's the best choice structurally for wide pieces.

One of the main techniques developed for dealing with movement on wide panels is called frame-and-panel construction. A frame of vertical and horizontal pieces is made, and a panel is set within it in a groove. The panel isn't glued in but floats in the groove so that it can expand and contract with moisture variations. Many cabinet and architectural doors are made this way.

Figure 1. Each piece of wood in a wide table top will cup as moisture in the air changes. But, by alternating the tree centers top to bottom on adjacent boards, the overall effect is reduced, as this exaggerated drawing shows.

Manufactured Panels

Many different types of plywood, fiberboard, and other manufactured panels (sheet goods) are now available. The advantages to working with these are stability and efficiency; the disadvantages are appearance, cost, and construction needs. Plywood is a very stable material, it won't change dimension or warp like solid wood will. It's also very time efficient to use because it's ready to cut up and glue together. However, plywood doesn't look like solid wood, even with a beautiful veneer on it. It looks like what it is: a manufactured panel. In addition, you must use a design that hides the ugly layered edge of the material. High-quality sheet goods are expensive, but this can be offset by the overall time efficiency of using them. Your design possibilities are limited with sheet goods because of that ugly edge and because you can't fasten plywood with the same variety of joints as you can solid wood. It is weaker in small areas, being prone to split, but when glued or screwed over a broad area, it has great strength.

Wood Glues

One of the main advantages of this very strong glue is that you can soften it with steam and disassemble a joint or joints for repairs (crucial to an antiques restorer). But hide glue is difficult to use because you must keep it hot in a special pot, and it hardens so fast that you have to assemble the work very quickly. Hide glue is no longer the most commonly used wood glue, although some manufacturers now market brands in a bottle that don't require heating in a pot.

Did you know???

Traditionally, wood glue was hide glue, which was made from the hides of cattle.

The yellow woodworker's glue you find in every hardware store is called aliphatic resin glue, a polyvinyl formulation. This glue is probably the most commonly used wood glue today. It sets up fairly quickly, so you can take the work out of clamps in an hour or two, and it cleans up easily with water. Most of these formulations are not waterproof, but they are water resistant, meaning that the joints won't come apart if you spill water on them. But don't leave the work outside in the rain unless you use a formulation that claims to be waterproof.

Urea formaldehyde glue, often called plastic resin glue, is a brown powder that you mix with water. Unlike aliphatic glues, this glue has a long open time, meaning that it won't set up quickly, so you can take a long time to put together a complicated piece, such as a chest of drawers carcass. This glue is more water resistant than aliphatic glues, and I've seen it used in boat building with success. Use it with good ventilation because of the formaldehyde.

The glues mentioned here are the most commonly used, and all are stronger than the wood surrounding them when used properly. These glues work best when they are just a thin layer between two well-fitting pieces of wood and aren't as strong when they have to fill gaps. The following are other glues that have special applications.

Resorcinol is the choice when you need a waterproof glue. This is a two-part glue that you mix together without water. It's tough as nails, but it leaves an obvious glue line.

Polyurethane glue is the new kid on the block. It's very strong, and it sets up fairly quickly. One of its main advantages is that it sands easily when dry and takes stains well, so it has less chance of showing a glue line. It can be used to bond other materials to wood.

Cyanoacrylate glue dries instantly and so is good for repairing small cracks that appear at the last minute, when you least need them.

Epoxy is the ultimate in strength, but because most wood glues are stronger than the wood around them, this is not as important. However, epoxy is waterproof and very strong when filling gaps. Cleanup is very difficult, requiring the use of nasty solvents.

Contact cement is the least strong of all wood glues, but it has one big advantage: It bonds on contact, making the application of veneers or laminates a snap. However, don't try to use this for structural woodworking applications.

Tools

New or Used?

You'll find many ads for new woodworking machines in the woodworking magazines, and who wouldn't want a nice shop full of new machines? If you can afford it, great, but few of us can. You can find used machines in the want ads of your newspaper if you wait long enough, but you need to be careful when buying used machines. Here are a few buying tips.

Generally, a used machine in good working order sells for something over half what it would cost to buy it new. Find out what the machine would cost new before you haggle the price. Remember that you need to base your judgment on the cost of a new, comparable machine. Say that you're looking at a table saw at a yard sale. The seller says that a new American table saw costs $1,200, so he'll sell you this one for $800. But you find out that this is a less expensive Taiwanese brand that sells for $750 new to begin with. I hope your neighbors are more trustworthy than this, or just more knowledgeable.

Inspect any used machine very carefully and operate it before you buy. Look for broken castings and loose, worn-out bearings. Check for wear and damage on all the little fittings that make the features of the tool work. On older machines, you might not be able to get replacement parts if some are missing or broken. On the other hand, some older machines were very well made and are a pleasure to own. Beware of the ubiquitous cry, "Well you can fix that!" Sure, you can hire a machinist to do anything, but you might end up paying him or her twice the cost of a new machine.

The cost of new machines varies widely. The dictum "You get what you pay for" applies here, and you'll pay for what you get. The least expensive machines work but not as well as better machines.

The quality of attachments is often lower, making accuracy more challenging. And service after the sale might not be what you expect.

Some years ago, the American market was flooded with very inexpensive machines from Taiwan, and many of these were of poor quality. But the Taiwanese responded, and now "Taiwanese" doesn't necessarily mean lower quality. Many American manufacturers have some of their machines made in whole or in part in Taiwan, Brazil, and elsewhere. What matters is how well it's made, not where it's made, and the only way to tell how well a machine works is to give it a try.

Hand Tools

The machine age has not entirely replaced the traditional hand tools of the past. Machines make most of the tasks that were done in the past with hand tools far easier and less time consuming, but not all of them. Skill with some hand-tool techniques will round out your vocabulary of woodworking abilities and save you time in some parts of the process. The following sections describe the most commonly used hand tools.

Planes

In the past, all rough lumber had to be straightened, flattened, thicknessed, and smoothed with hand planes. Now we can straighten, flatten, and thickness with jointers, planers, and table saws. But still the fastest way to smooth and finish a broad flat surface is with a sharp, well-tuned smoothing plane (see figure 2). With sanding, you must start with a rough grit and work through successive finer grits to get to the finished surface. A smoothing plane will take a rough surface to a finished surface that needs no sanding in one step. Small block planes come in very handy for knocking off a

Figure 2. Use a hand smoothing plane to quickly bring a rough board to a finished surface.

sharp corner or smoothing an edge. And they still do well what they were developed for, which is cleaning up end grain.

Chisels

In the past, a woodworker would have had several different sets of chisels, such as long heavy ones for chopping mortises and small light ones for bench work (small joinery cuts with the work clamped to your bench). Now we cut mortises with a drill press or a router, and chisels are necessary for squaring up the mortise. They are also irreplaceable for such tasks as hardware installation and making necessary adjustments.

Carving Chisels

These are altogether different from bench chisels. Carving hasn't changed much at all from its practice in the past, and a wide range of chisel shapes is still available for the carver. Some carvers have hundreds of chisels, and here's why: Carving chisels are available in countless shapes and sizes, and the more you have, the easier it is to do the work. The secret to carving is to have a tool that is shaped to make the shape of cut you want rather than, for example, trying to make a round shape with a straight chisel. Because of the many different shapes of cut that are possible, you can't have too many different shapes of chisel. You can also design your work around the shapes of the tools you have available. Many carvers get by with a limited number of chisels.

Saws

In the past, we had to cut boards out of the tree, rip them to width, and cut them to length, all with handsaws. Whew! Makes me tired just thinking about it. However, you'll still find many uses for your crosscut and ripsaw, and having sharp ones close by can be a big help. The handsaws used by woodworkers today are mainly the little ones, such as the dovetail saw. These have very fine teeth and are necessary if you choose to cut out dovetails by hand, but they also come in handy for making all sorts of little cuts here and there. Many woodworkers prefer the Japanese handsaws, which cut on the pull rather than the push stroke.

Scrapers

A scraper is a thin piece of high-quality steel that you put a burr onto and use to finish scrape a surface (see figure 3). Before sandpaper, this was the main tool of choice for finishing any surface that couldn't be planed smooth. Scrapers are still faster than sandpaper, but it takes some doing to learn how to get a good burr on the edge of the tool. Burring tools that make the task easier are available, and they come with thorough instructions. You will seriously reduce the amount of time you spend sanding if you learn to use a scraper.

Figure 3.   A scraper removes less wood than a plane and leaves a fine finished surface.

Machine Tools

Table Saw

For most shops the primary tool is the table saw (see figure 4), where most things begin and where many other procedures are done. A table saw is simply a table with a circular saw blade coming

Sharpening Stones and Grinder

The modern woodworker still needs to know how to get a sharp edge on a plane iron or a chisel. To do so, you need to start with a grinder. This tool removes a lot of metal from a tool edge quickly and is used to establish a bevel on the edge at the right angle or to remove nicks before sharpening.

You need only two stones to sharpen chisels and plane irons: a medium and a fine stone. The most economical and fastest working stones are the Japanese water stones, although traditional Arkansas soft (medium) and hard (fine) stones work well, too. One medium stone (800 to 1,200 grit) is all you need to smooth the rough surface left by the grinder and get the edge ready for final sharpening. The fine stone (6,000 grit) is so fine that it can only smooth the surface left by a medium stone. It won't level the Grand Canyon–depth gouges left in the edge by your grinder.

Use a sharpening jig to make the task easy. This tool holds your chisel or plane iron on the sharpening stone at a steady angle while you push it across. Grind the tool to about a 25-degree angle, then put it on the medium stone with the jig so that the bevel is almost parallel to the stone but with just the tip touching. Sharpen until $1/16$ to $1/8$ inch of the bevel tip is smoothed. Then change the position of the tool in the jig so that it hits the stone at just a slightly steeper angle. This ensures that as you sharpen on the fine stone, only the tip is touching, which is the only surface you need to polish. Now work the tool on the fine stone just until the tip is polished all the way across. You also need to polish the flat side of the chisel or iron by polishing flat on the stone.

up out of the middle and with a fence parallel to the blade to guide the work. Use a table saw to rip parts to width, cut shorter ones to length, and cut long dadoes (or grooves) or rabbets (a groove on the edge of a piece). Numerous attachments can be made or bought for the table saw for making special cuts. One attachment that comes with all new table saws is a miter gauge. This slides in a groove in the top of the table next to the blade and has an adjustable fence on it that allows you to make angle cuts on the ends of boards.

The two basic types of table saw are tilting arbor and tilting table. The tilting arbor is by far the more common. On the tilting

Figure 4.   A typical 10-inch table saw with rip fence and miter gauge.

arbor machine, the table stays put while you turn a crank to cause the blade to tilt from 90 to 45 degrees for making angle cuts on the edge of boards. On a tilting table machine, the blade stays put while you turn a crank to tilt the table from horizontal to 45 degrees for making angle cuts. Tilting arbor machines are generally easier to use, although some tilting table machines have a sliding table, where half the table on one side of the blade slides on rollers so that you can clamp work to the table and slide it into the cut. This can produce accurate miter cuts more easily than can a miter gauge.

The capacity of table saws is a function mainly of blade size. Ten-inch blades are the most common size for small shops. Smaller saws are made with 8- or 9-inch blades, but these tools have small motors on them and are not useful for cutting wood much over ¾ inch thick. This might be all you need, though, depending on what you're going to do. Most 10-inch saws have cast-iron tables that are roughly 30 inches square, and most woodworkers build plywood extension tables onto their saws to increase the overall table size.

Table saw fences come in a variety of geometric configurations, but the fences that come with the machines are often lacking in performance. You want to be able to quickly move a table saw fence toward or away from the blade and have it stay parallel to the blade in any position for a smooth cut. Numerous bolt-on fences that are well designed and a joy to use are available for table saws. In addition, these fences give you greater capacity by extending the fence as much as 4 feet from the blade for cutting sheets of plywood.

Handy Hint

Motor horsepower is an important factor to consider when buying a table saw. A ¾ horsepower motor just doesn't have enough guts to cut lumber much more than ¾ inch thick. Two horsepower is nice, but you'll do okay with 1½.

Band Saw

This tool has two (sometimes three) wheels with a thin metal blade that runs around them and down through a table (see figure 5). Because the blade is thin, it's easy to cut curves. I know a hand-tool woodworker who has only one machine tool in his shop: a band saw. He uses it to rough out his lumber and cut curves, then he cleans up the pieces with his hand tools. A band saw is similar to a table saw with respect to its versatility, and if it isn't the first machine you get, perhaps it should be the second. Use a band saw to cut curved shapes, rip parts to width, and resaw to thickness. Resawing is splitting a board down its thickness, making two 1-inch-thick boards out of one 2-inch-thick board. When you have big gnarly chunks of found wood or locally milled lumber, a band saw is useful for trimming them in preparation for other procedures. You

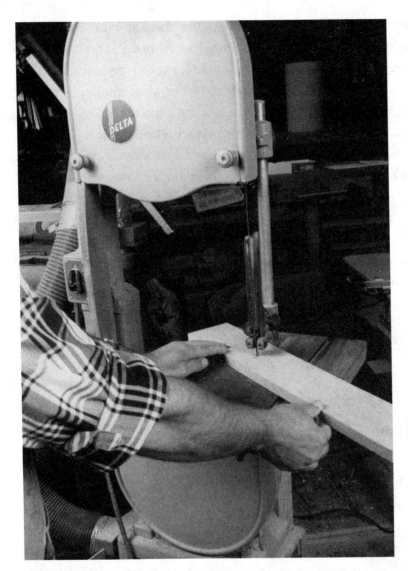

Figure 5.   Use a band saw to cut curved shapes in wood.

can do a number of special setups on a band saw, including cutting dovetails and cutting tenons.

Most band saws are of the two-wheel type, but the three-wheelers will give you a greater throat depth. Throat depth is the maximum width you can cut between the blade and the upright arm of the saw that holds the upper wheel(s). Although three-wheelers give you a greater throat depth on a less expensive machine, they tend to break blades faster because the wheels are smaller. A smaller wheel causes the band to bend more each time it goes by, stressing the weld and band steel more than a larger wheel does.

The most common two-wheel band saw sizes for small shops are 9, 12, and 14 inches; these numbers are measures of the wheel and throat. Nine-inch machines are suitable only for the smallest work. Along with a band saw's throat size, look at its height of cut, which is about 6 inches on 12- and 14-inch machines; you can find extension kits for some band saws that raise the upper wheel another 6 inches to give a cutting height of 12 inches. This capacity is very important for resawing wide boards.

You'll get by with a ⅓- or ½-horsepower motor on a band saw, but you'll want a ¾- or 1-horsepower motor if you plan to make many thick cuts, such as roughing out thick chunks or wide resawing.

Radial Arm Saw

This is a motor and circular saw blade mounted above a table on an arm with little wheels that allow you to pull the blade across stationary boards on the table (see figure 6). Although it is possible to make a variety of different cuts with a radial arm saw, the reality is that it's more difficult to make most of these cuts with a radial arm than with other tools. A radial arm saw is no substitute for a table saw. However, a radial arm does at least one thing better than a table saw: making crosscuts. Most shops set up a radial arm saw with a long extension table to the right and/or the left so that you can slap a long board up on it and cut it off quickly and accurately. In addition, you can mount a dado cutter onto the arbor and cut dadoes across the grain. It's easier to do these things on longer boards on a radial arm than a table saw because on a table saw you have to move the work across the tabletop during the cut, which is difficult to do with long pieces.

Figure 6. A radial arm saw makes crosscutting long boards very easy.

Some very small radial arm saws have been made that take 8- or 9-inch blades, but most take a 10-inch one. You'll find a lot of these in the want ads. You might run across a larger one here or there, but for a small shop they're overkill. Save your money for a better table or band saw before you spend a lot on a radial arm. Many small-production cabinet shops use the inexpensive 10-inch variety and do just fine. Larger ones are for big shops that cut a lot of thick wood.

Jointer

A jointer is a very simple tool that cuts a straight line. It has two tables with a spinning cutter head between them, and you push the work from one table to the other across the cutter head, trimming the edge or face (see figure 7). Use it to quickly straighten one edge of a board, then straighten the other on the table saw, running the jointed edge against the table saw fence. You can also use a jointer to flatten the face of a board, in which case you'll need a planer to bring it to uniform thickness again.

Jointers are critical in large shops that use a lot of solid wood in production setups. Cabinet shops that use sheet goods and surfaced 1x can get by without one. But to do production setups on large quantities of wood accurately, the wood must be straight, and for that you need a jointer. Wood, as it comes off the truck, will not be straight and as such cannot be used in precise machine setups. If you don't plan to do a lot of production work, you'll find a jointer very handy for straightening the edges of boards and other things. But for a small shop, it might be a luxury, and you should get a good table saw first.

The main jointer capacity is width of cut, which determines how wide a board you can face joint. Four-inch machines are useful only for the smallest work. You can squeak by with a 6-inch machine for furniture work, but you'll be happier with an 8-inch one. Some machines don't have adjustable out-feed tables. This makes it much more difficult to adjust the machine to get an accurate, straight cut. If you can, spend the extra money to get a machine with an adjustable out-feed table.

> ## Handy Hint
>
> Keep a can of paste floor wax in the shop and rub it into your machine tables from time to time. It makes wood slide over the table easily and reduces rusting of the iron.

Planer

This machine evenly reduces the thickness of pieces of wood (see figure 8). Overhead it has a spinning cutter head; rollers push the

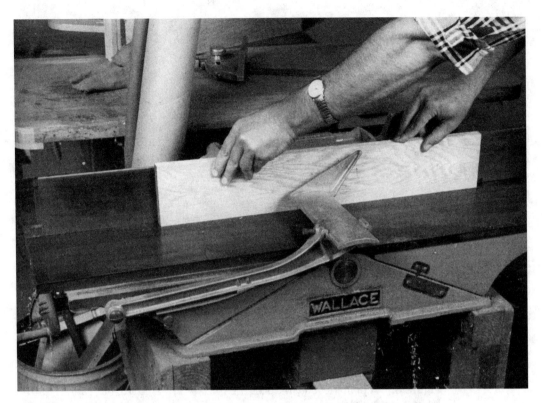

Figure 7. The basic purpose of a jointer is to make a straight line or surface on the edge or face of a board.

board by this cutter head. The machine keeps the other face of the board a constant distance from the cutter head, ensuring that the board comes out at a uniform thickness (which a jointer can't do, but a planer can't straighten). You can fashion jigs to help a planer make its parallel cut on a bevel or a taper, but this is rarely done. Most people use a planer for thicknessing, and there's no other easy way to bring your wood to a lesser thickness than how you bought it. Until recently, even smaller planers were expensive, but now a variety of new machines are available for about $400. These machines work, but they have limited power and can't remove much wood at one time. Still, for the small shop with a limited need for a planer, they're a viable alternative.

Figure 8.   A planer evenly reduces the thickness of a board, but cannot flatten it like a jointer can.

Width of cut is the main capacity issue with a planer, and most will cut a board at 12 inches wide. The next tier of machines costs at least twice as much as the previously mentioned ones, and many are only 12-inch machines as well. However, their more powerful motors will take a deeper depth of cut, greatly reducing the time it takes to finish the work. These machines are also heavier and will stand up to more use. Get a more expensive machine if you need to do a lot of planing.

Router

Here's one of the most versatile machine tools for the small wood-shop. It's unequaled for the number of applications and variety of

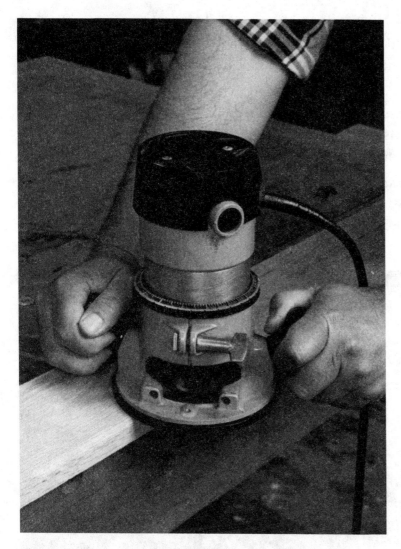

Figure 9.   Use a router hand-held to cut shapes into the edge of boards.

cut shapes possible. It's simply a small motor with a fitting on the shaft, called a collet, into which you install a little cutting tool, called a bit. The motor has a base on it with handles so that you can hold the machine or screw the base to a table (see figures 9 and 10). The following is a brief outline of common router operations.

Edge details: Put a bearing guided bit in the router, such as a round-over bit, and run the tool along the edge of your wood to round it over. Many different shapes are available for different edge detail looks.

Joinery setups: Build a jig to guide your plunge router to cut mortises in the edge of frame pieces. In a router table, use stile and rail bits to join the frame parts on cabinet doors.

Dovetail jigs: Numerous jigs are available for cutting dovetails with a router. Some are simple and do one thing, and others more complex and cut a wide variety of dovetail joints.

Figure 10.   You can also place a router upside down in a table as shown and then push your work against the router bit.

Router table: Put the router in a table with the bit poking up through a hole in the middle, and you have a miniature shaper. Router tables have been used for more setups than you can imagine, from edge details to template shaping and from joinery setups to panel raising.

Template shaping: Make a curved template with good plywood (no voids) and use a bearing-guided flush trim bit to cut out pieces of wood at the same shape as the template. Attach the template to the wood, cut off the bulk of the waste at the band saw, and then flush trim on the router table or with the router handheld.

Did you know???

The router was originally a hand tool. Used by coachmakers and other woodworkers, the hand router is similar to a spokeshave but cuts a molded shape like a molding plane. They were devised for cutting shapes along curved surfaces.

Special jigs: If you can imagine how a certain router bit can make the shape of cut you need, then imagine how to build a jig that will guide the router reliably along the cut. The applications are limitless.

Shaper

This is the large machine tool that a router table replaces. A shaper is a cast-iron table with a spindle coming through a hole in the middle on which you mount any of a wide variety of cutters for a number of applications. You should consider using a shaper only if you have a lot of large production work to do it or if you are doing special work, such as making large curves to templates, where the capacity of a router is inadequate. A shaper is one of the most potentially dangerous of woodworking machines, so you should always use it with great caution.

Figure 11.   Bore accurate, well-aligned holes with a drill press.

Drill Press

A drill press makes boring accurate, well-aligned holes fast and simple (see figure 11). A chisel mortising attachment to a drill press lets you bore square holes in wood for mortises. You can also mount small sanding drums in it for sanding curved edges. Many small bench-mounted drill presses are available, and these often are adequate for the needs of a small shop.

Boring Machine

This machine is made for doing dowel joinery. It punches holes horizontally rather than vertically, like a drill press does, so it's easy to set up for boring dowel holes in the edges and ends of pieces for joining table or chair legs to rails or for making cabinet or architectural doors. This is a production tool dedicated to large runs of doweled components, and you should consider using it only if you plan to do a lot of this. Otherwise, you can get by with simple dowel jigs and/or a drill press.

Lathe

This tool spins the work around while you gently push a cutting tool into and along the spinning surface (see figure 12). It is used for making round table legs and so on. Lathes come in all sizes. Swing and length are the two main capacities to consider. Swing refers to the distance from turning center to the bed and determines the diameter of the piece you can chuck up into the tool. Length is how long a piece will fit in it. Very small lathes are available for making miniatures; if you are ambitious, you can get a 15-foot architectural lathe and make columns for Greek revival remodels. Most lathes are long enough to turn a table leg and have enough swing to make a large salad bowl. For turning legs, candlesticks, and other spindles, you mount the wood between centers so that it's spinning on both the headstock (where the drive pulley is) and the tailstock (which moves back and forth to adjust for length). To turn bowls, the wood is attached only to the headstock, where it is screwed to a special faceplate or held tight by any of a number of types of chucks, which are metal devices that grab the wood and spin with it.

The tools that you use to cut with are specialized for the lathe and come in a variety of shapes and sizes. Gouges have a U-shaped shank and curved tip and are used for rough work and finished curved surfaces. Skews have a straight edge set on an angle and are used for detailed finish work. The shape of your cutting tool determines what it is useful for, much like in carving. Specialized tools are available from small specialty manufacturers, such as a curved tool for cutting nested bowl blanks out of a single chunk of wood to minimize waste.

Figure 12. Use a lathe to spin a piece of wood so that you can shape it into a cylinder as shown, or make more elaborate forms.

Figure 13. Use a stationary belt sander to sand smooth surfaces.

Sanders

Many different mechanical configurations of sanders are available, and each design lends itself to certain applications. Here are the major ones:

Stationary belt: A bench-mounted machine, this has a belt, the face of which moves vertically (see figure 13). It has a small horizontal table at the belt that you can place work on as you push it into the belt. This is one of the most versatile types of sanders.

Disk: Often included with a stationary belt sander, a disk sander is simply a large round flat plate with a round piece of sandpaper glued to it. It has a table like the stationary belt.

Hand-operated belt: These handheld machines have a sanding belt on the bottom surface and are famous for use in surfacing large, broad areas but infamous for leaving divots if you tilt it the wrong direction (see figure 14).

Figure 14.   A hand belt sander is best used for smoothing large, flat surfaces.

Orbital: These small handheld machines are used for finishing smaller surfaces (see figure 15). A recent innovation is the random orbital sander, with a double-action head that causes the tool to work faster.

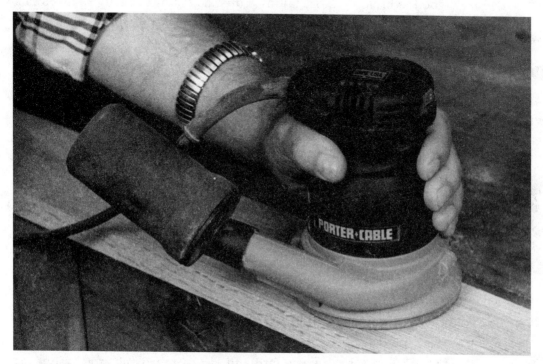

Figure 15.   Do finish sanding with a hand orbital sander, using finer grits such as 120, 180, and 220.

Scroll Saw

Often called a jigsaw, this tool has a thin reciprocating blade that sticks up through a small table for cutting sharp curves in thin pieces of wood, as with jigsaw puzzles. Fretwork consists of panels with intricate designs in them cut out with a scroll saw. These panels are then incorporated into larger pieces of furniture or other objects.

Dust Collector

This is a large vacuum machine with pipes, or ducting, leading to your machines to pull dust away as it is produced. If you use machines fairly often, you'll produce a lot of wood dust, which is not good to breathe in. Installing a dust collector not only reduces this threat but also makes cleanup a lot easier. A 1-horsepower dust collector is inexpensive and will handle pulling dust from one machine at a time as long as the ducting is not too long, and this is the case in most small shops. A dust collector is a worthwhile investment.

Did you know???

There are few things you can do with a belt sander that you can't do just as well or better with a sharp hand plane or a scraper. You can, however, have races with belt sanders. That's right, people actually gather together to race these mechanical demons down special tracks, dragging extension cords all the way.

Jigs and Multitools

A wide variety of jigs and multitools are available for doing various operations. A jig is simply any device that helps you use a tool to do a certain operation better. Dovetail jigs help you use a router to cut dovetails by aligning the boards and individual cuts for an easy, precise dovetail joint. The distinction between a "jig" and an "attachment" is not very hard and fast. A mortising attachment for your drill press attaches to the drill press to let you cut square holes. Is it really a jig? You decide.

Multitools, however, are distinct from other jigs and attachments because they are designed for versatility. They are elaborate router jigs with a variety of tables, fences, and other attachments that let you easily set up for joinery operations, such as mortise and tenon cutting. These devices accomplish what shop-made jigs or other jigs made for specific purposes can, but multitools do a range of things with one device and are quick and easy to set up precisely. However, they are costly and are most useful for production setups.

Combination Machines

Several manufacturers offer machines that do the operations of several different wood-working machines in one tool. The Shopsmith basic tool has a table saw, lathe, drill press, and disk sander in one package, and you can get attachments, such as a band saw, jointer, and planer. The advantages of such machines are that you get more tools for the money, and it takes less space than three or four separate machines. Disadvantages are that smaller combination machines have smaller tables and are not as convenient to use as larger, separate machines, and you must change the setup on the machine for each operation. You must disassemble the table saw to use the drill press and vice versa, whereas with separate machines you can leave one set up as it is and use another. But many hobbyists choose combination machines because of space limitations.

Electric Hand Drills

These units have greatly improved since the days of the little aluminum-bodied single-speed hand drill that smelled like an electric train when you used it. Battery-powered hand drills are the order of the day now, and they are very convenient although still more expensive than the cord variety. Be sure to get a hand drill with a variable-speed trigger so that you can put a screwdriver tip in it and gently ease screws down into wood. I have two drills: a single-speed cheapy with a cord for drilling holes and a good battery drill with a variable trigger to sink the screws.

Biscuit Joiner

This handheld electric tool is a more recent innovation and makes much joinery far easier (see figure 16). It simply cuts a little slot in the edge of a piece of wood. Then you put a special wood spline in the slot, called a biscuit (these are supplied by various manufacturers in uniform sizes). The tool aligns the slots easily so that you can quickly glue boards together on edge and keep the faces aligned to each other. These are useful for many other joinery operations as well, such as joining solid wood to the edge of plywood or even joining chair legs to their rails.

Figure 16.   Use a biscuit joiner ro cut small slots in the edge of boards for splines that will join two boards together.

Clamps

The main thing about clamps is that you can never have enough. Every woodworker eventually gets into a situation in which he or she wishes there were just a few more. Gluing things together with many clamps, each with a little pressure, is far better than using a few clamps with a lot of pressure. When you see a clamp at a yard sale, buy it.

Many different types of clamps are available, and each suits a different situation. Large bar clamps apply great pressure over a long distance, whereas C-clamps apply great pressure over a short distance. Lighter bar clamps apply lighter pressure over a long distance, whereas hand screws apply lighter pressure over a short distance. In some situations, specialized clamps, such as a strap clamp (which is a long nylon strap with a small winch on it), work great. Use this to pull together four legs of a chair or table all at once rather than putting separate clamps on each side.

Obtaining Materials

Where is the best place to get your lumber and tools? Investigate the following alternatives and choose the best source for you.

Most lumberyards cater to the building trade. Although they have some hardwoods, they might not have much of a selection. Stores that specialize in hardwoods will have more types and a better selection of sizes in a given type. Some of the mail-order woodworking catalogs offer dimensioned lumber, but this is expensive. The classifieds ads of woodworking magazines might list businesses that will ship lumber to you. Try small orders from such companies until you trust that what they send you is what they described over the phone. Before you buy, ask a lot of questions about grade, sizes, cut (quarter sawn or flat sawn), and shipping fees.

Some lumberyards and hardware stores sell new machines. You can always get the same machine cheaper by mail order, but remember that you won't be able to look at it first, and you won't get the tool right away. A wide variety of woodworking tool catalogs are available, and as soon as you get your name on some magazine mailing lists, you'll receive many of these in the mail. You can also order catalogs directly from the ads. Each catalog will carry one or two brands of machine tool, offering all the different types and models offered by that manufacturer. Shop around. You'll find wide variations in price for the same machine in different catalogs.

Some catalogs are general and cover a wide array of woodworking supplies, from small tools to large machines and from commercial tools to hobbyist tools. They will also have sections on hardware, finishes, and so on. Some catalogs are dedicated to only one type of supply, such as hardware, machines, or turning tools. These catalogs will give you more choices when looking for a specific item. The resource section at the end of this book lists many popular catalogs.

Other Supplies

Countless other supplies are available for woodworking. Remember that once you get your name on a magazine's mailing list, you'll start receiving catalogs offering any supply you might need. Many of these supplies are specialized tools that are useful for one specific operation. Before you buy any specialized tools, think about how much you'll really use them or whether you would be better off finding your own solution.

Two larger areas of woodworking supplies are hardware and finishes, so let's take a brief look at how to find these.

Hardware

The trick with hardware is finding just the right thing. You can custom design your wood, but customizing hardware is far more difficult. It's best to use what's available. You might find good furniture hardware in your local hardware store, but the dedicated woodworking hardware catalogs are the best place to look because of the wider variety they offer. Most catalogs carry at least some hardware.

Finishes

Like hardware, you'll find some finishes in the local stores, and you might find just what you need there. Polyurethane varnishes and wipe-on oil finishes can be bought off the shelf, but for more esoteric finish work, such as that requiring shellac or top-quality tung oil varnishes, you're better off looking in catalogs for the best materials.

Setting Up Your Personal Workspace

A SHOP IS SIMPLY THE SPACE that you dedicate to practice your craft in, and it needs to accommodate the types of projects you'll be working on. Your shop need not be an industrial installation; it could be simply a corner of the kitchen. I once met a man who made miniature wooden ship models on his kitchen table using just a few small tools. Regardless of the type and scope of your woodworking, this chapter will examine some things you should consider when setting up a shop.

Your woodshop will be a place of refuge for you. To the extent that you can, make it as pleasant a place as possible. I'm fairly lucky; my shop is in an old livery stable with big windows, skylights, and pleasant surroundings, although it's cold in the winter. Think about lighting, ventilation, heating, and cooling as you look for a shop space. If you are limited to a space that doesn't have optimal conditions, such as a small garage or other outbuilding, you might want to invest in another door or window. As a bonus, this will allow you to just gaze outside and daydream if you want.

If you plan to get a lot of machines, think about your electrical facilities. You might need to put in a few more circuits and maybe even a new panel with breakers just for your tools. Many wood-working tool motors can be wired for either 110 or 220 volts.

Be sure to get a fire extinguisher for your shop and keep the shop as clean as you can to avoid fire danger. Accumulated dust, shavings, and chips and chunks of wood in the corners or behind things are little piles of kindling just waiting for a fool with a cigarette or sparks from a failing machine to ignite them. Remember that a pile of oily rags can spontaneously combust.

Handy Hint

If you can, wiring for 220 volts will reduce your electric bill and cause less strain on the motors. If you're an electrical dummy like me, call a pro and get it done right. If you do it yourself, remember to always install a ground wire on all circuits for safety and hide wiring either inside the wall or inside conduit, as exposed wiring can be hazardous.

Organizing Your Woodshop

How you arrange your shop will be a function of what machines and benches you have and how much space you have. However, a few generalizations apply in all cases, so let's take a look at those. With each machine (and bench) that you have, look carefully at how long pieces of wood will hang over the tool during use. For example, on a table saw you push wood along the board's entire length. Thus, to rip an 8-foot board, you need 8 feet behind the saw to start and 8 feet ahead to finish. That's 16 feet, plus a little extra needed to make such a cut. For each machine, you want to maximize the usable distances that are available, but doing this so that wood coming out of one machine doesn't bump into another can be a challenge. One trick is to set up machines so that the paths of wood traveling across them are parallel. Thus, a band saw can go right next to a table saw without either interfering with the other.

In a very small shop, I once saw a planer right up against the wall. I thought that the planer must have been in storage, as any wood put through it would have hit the wall on the way out. Then I saw that a 1-square-foot door had been cut through the wall so that boards could pass through it as they were planed. Other solutions to problems of limited space are to locate machines near doors so that work can pass through the doorway or placing machines on rollers so that they can easily be moved around. You can buy dollies that are made just for that purpose, or you can make your own. If you make your own, get locking wheels so that you can anchor the machine while it's in use.

Workbenches

What kind of workbench should you get? Although traditional workbenches with wooden clamps and trestle legs look impressive, these might not be practical, depending on the type of woodworking you do. Traditional benches were designed for the needs of a traditional hand-tool woodworker, who takes individual pieces of wood, clamps them into the bench, and then does a variety of operations on them. For this you need a rigid bench with as many clamping options as possible. In machine-tool woodworking, more often you are pushing the wood through a machine rather than clamping down the wood and cutting it while clamped. For this reason, you have less need for a traditional bench.

I have two benches, both of which I made. The first is traditional, and I still find it very handy, even though most of my woodworking is with machine tools. When I need to do hand-tool work for installing hardware, planing, and so on, the traditional clamping techniques can't be beat. My second bench consists of two cabinets of drawers with a benchtop mounted above them. All those drawers come in handy for storage, and the perimeter of the benchtop (which is 1½-inch-thick laminated oak) is unobstructed by built-in vises, so it's easy to clamp machine jigs to it. Invest the money to buy or the time to make a traditional bench only if you plan to use its capabilities by learning hand-tool techniques.

A variety of benches are available by mail order through woodworking catalogs, ranging in price from a few hundred dollars to much more. Look carefully at the size, quality, and features of any bench you consider buying, and be sure to ask before you buy whether you can return it if you don't find it satisfactory. You might have to pay shipping in such a case, but this might be worth it in order to get a good bench.

Efficient Ducting

If you decide to invest in a dust collector, be careful about how you install it so that it works correctly. Nothing is worse than ducting that clogs up all the time. The manufacturers usually have useful booklets that tell you how to install the ducting correctly. Regardless, the basics of ducting are fairly simple in that you want a smooth airflow over the shortest distance possible.

Your collector pulls air through the ducting, but it must fight the resistance that is caused by bends in the ducting. The more sharp the bend, the greater the resistance. Don't use 90-degree T-joints to join two pieces of duct. One pipe should intersect another at an angle, but be sure to design your system in a way that minimizes the number of turns that the air must take.

However, even straight ducting causes resistance to airflow. A 1-horsepower collector can't be expected to reliably pull dust and chips through more than 25 feet of ducting from machine to collector. Thus, you should place those machines that require the most dust collecting as close to each other and as close to the collector as possible. These machines are, in order of need for dust collecting: the table saw, the radial arm

saw, the planer, and the jointer. Stationary sanders and band saws
that are used a lot should be ducted as well.

All the ducting and attachments you need for the collector can
be expensive, so improvise. I used 4-inch cardboard carpet tubing in
place of metal ducting, and I made my own blast gates (the valve at
each machine) and Y-joints. However, you can't make flex tubing
yourself. Luckily, some types are inexpensive and readily available
through the catalogs. Flex tubing is very helpful for connecting the
ducting to individual machines.

Do You Need a Spray Booth?

If you turn your hobby into a business, or the business of your hobby turns out
to be very productive, you might consider using a spray gun to do your finish work.
Nitrocellulose lacquer (just called lacquer) was developed for the sake of providing a
fast way to spray-finish wood. Water-based lacquers are increasingly being used today.
Regardless of what you spray, you need to have a spray booth if you do a significant
amount of spraying.

A spray booth consists of a wall of filters through which air is drawn by a powerful
fan. You place your work close to the filters in an enclosed booth and spray there.
The fan pulls the overspray away and through the filters before expelling the vapors
out. Without this means of eliminating overspray, you would fill your shop with a
cloud of mist.

Few hobbyists will invest in a spray booth, but if you do a large volume of work, you'll
find that spraying greatly reduces finish time, so you'll want to use one. You can visit
your local cabinet shops and try to find a nice person who will let you use theirs, but
don't ask to use their spray gun. These tools must be carefully cleaned after use, and
they don't want you to be responsible for that. Find a used spray gun at a flea market
or invest in a new one. However, you'll need to use the cabinet shop's compressor, so be
sure that you know how to turn it on and off and bleed out condensed water from the
compressed air.

Closing Thoughts

The most important thing about setting up your shop space is customizing it to your needs. Your shop is a specialized tool in itself that is designed for a certain purpose—the pursuit of your craft—so don't hesitate to arrange or alter your personal space in any way that will help you do that. Be it half a garage, an entire barn, or a corner of a room in an urban high-rise apartment building, make your shop *your* shop!

Creative
How-To's

YOUR TWO MAIN GOALS when designing a piece of wood-working are appearance and function. It's not a good idea to sacrifice one for the other, for example, making something that looks wonderful but that either won't be useful as the thing it's intended to be or will break or fall apart when used. On the other hand, you could make something that is very strong and very functional, but with such odd proportions or unpleasing design that no one wants to have it in their house. Because with most woodworking it's difficult to make radical alterations to a piece once you have begun to build it, you're better off carefully planning and even building a mockup or some mockup parts before you begin to make sure that you're on the right track.

I always begin my design process with a few pencil sketches. At first, I'm looking only for a general sense of the proportions I want to use in the piece. The most important part of dealing with proportions is not to come up with an absolutely perfect relationship between, for example, width and height but to avoid coming up with a relationship between the two that's really unpleasing. Of course, this is a very subjective issue, and what is pleasing to one person is another person's anathema. In general, however, things that are far higher than they are wide look skinny, and things that are roughly the same height as their width look squat and boxlike. A good rule of thumb for making pleasing rectangular shapes is to make roughly a 1-to-1⅓ relationship between the two sides.

However, don't feel like you're stuck using only this formula. The front of a desk might need to be twice as wide as it is high, or perhaps you like the idea of a tall, skinny box with a whole bunch of little drawers stacked up one on top of the other. But the side of the desk can have the 1-to-1⅓ proportion, as could the drawers in the box. At least try these proportions in various parts of your pieces as you sketch out ideas. The good thing about sketches is that they take little time and can be easily altered or relegated to the circular file.

Handy Hint

When marking your wood, use a scribe rather than a pencil. The line of a pencil in wider and therefore less accurate. To make a scribe, grind a nail or small screwdriver to a tapered pin point.

Once I've come up with a good basic sketch of what I want to make, I look at the basic structural design for the piece. Do you want to make it with frame-and-panel construction, or do you want solid sides? These are the two basic construction methods for most furniture designs. Frame-and-panel work consists of making rectangular frames with horizontal and vertical pieces of wood joined at the four corners and then placing a panel inside each frame. You then join the frames together to make a carcass. Solid construction simply means that the sides of the piece are solid, wide pieces of wood, as in the case of a hope chest made with wide pine boards (a common design in colonial furniture).

At this point in my design process, I look seriously at the possibility of harmful wood movement. If I want to make a piece with large, wide pieces of wood for the sides, what's going to happen when these pieces expand and contract with seasonal moisture variations? Because frame-and-panel construction avoids this negative effect, you might choose it just for that reason, but if you want to use solid construction, you simply need to design the piece such that it can expand and contract without breaking itself. This means attaching wide tabletops with slotted keepers that can slide and, in general, avoiding gluing or otherwise permanently fixing pieces of wood at right angles to the grain over more than a few inches. There is always a way to let the wood move, and you must let it move because, believe me, it will.

Once I've decided on my basic proportions and construction method, I get down to business with a scale drawing. For this get a draftsperson's scale, which lets you draw at scales of ¼ inch to the inch or ½ inch to the inch. These tools, which are available at stationers stores, are simply rulers that let you measure at these smaller scales. This is convenient because you don't have to make a full-scale drawing. The scale drawing makes you come up with specific dimensions that manifest your proportions and makes you look at the joinery carefully. Make section drawings through the piece. This means drawing it as though you took a big knife and cut the piece into two sections and you are looking at the sliced section surface. This kind of drawing forces you to show where things join and are located with respect to one another. You don't want to find out halfway through construction that things don't line up, so make sure they do now.

If my piece has curved parts in it, I make a full-scale drawing of it on a large piece of paper or on a piece of plywood. This drawing will be necessary later for positioning the curved parts in relation to the other parts.

If my piece is complicated or I am concerned about getting the proportions right, I build a mockup of the whole piece. Cardboard is a good material for this, as are scraps of plywood. The point with a mockup is to see a three-dimensional object instead of a bunch of little drawings and visions dancing in your head. This physical demonstration of your idea is the best way to make adjustments and to notice major blunders that you hadn't thought of.

Then, once I'm sure of my larger design, I look at how I will do the joinery. By this point, I've already decided on what the major joints will be, but now is the time to look specifically at those and all the other joints in the piece. The kind of joinery you use will be a factor in determining the exact lengths of many of the pieces of wood you make, so it helps to make full-scale drawings of joints to show mortises and tenons, dowels, dovetails, splines, and so on. Once you've determined these things, you can make a cut list and start looking for lumber.

Construction Specifics

Following are the major types of woodworking joints and some of their uses (see figure 17). Later in this chapter, we'll take a closer look at which tools to use in making these.

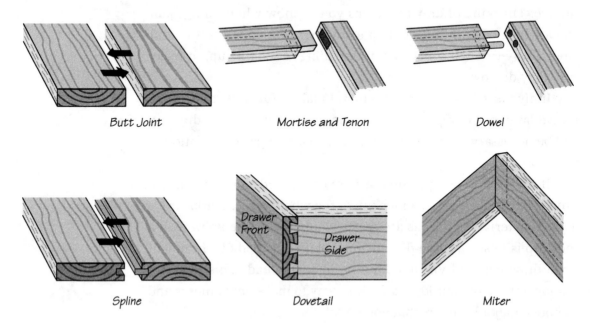

Figure 17. Basic woodworking joints.

Butt Joint

A butt joint is simply where two pieces of wood butt up against each other with no tenon, dowel, spline, or other joint feature between the two. You can't use this joint on the frame parts of a cabinet door, but you can use it where the joint has a broad area for a good glue bond, such as in edge-gluing pieces together to make the wide panel that goes into a cabinet door frame.

Mortise and Tenon

The mortise and tenon is the traditional joint of choice for joining frame components. It consists of a hole, or mortise, cut in the long edge of one component and a finger, or tenon, cut out of the end of another. The tenon is glued or pinned into the mortise to hold it fast.

Dowels

Recently, the dowel joint has replaced the mortise and tenon for much frame construction because it is faster to make. Holes are bored into both mating parts, and a round peg or dowel is glued and located between the two. Although this joint is not as strong or long-lasting as the mortise and tenon, it is still more than adequate for most work.

Splines

A spline is a thin strip of wood placed into a groove that is cut in two joining pieces of wood. Locating a spline in the butt joints of boards being glued together for a tabletop or other wide panel helps keep the surfaces of the boards aligned during gluing. When a spline is used structurally like a tenon, locate the grain direction perpendicular to the joint, or the spline will split when stressed. Biscuit joints are a kind of spline joint.

WELLES'S STORY

Artist and craftsperson Welles Goodrich of Bonny Doon, California, has approached woodworking both from a practical and artistic perspective. After seeing the devastating effects of the Vietnam War in a military hospital here in the states, he decided that he wanted to pursue happiness by pursuing beauty. Along the way he has done a wide range of woodworking such as woodcarving, new house construction, store improvements, cabinets, furniture, boat remodeling, and art furniture.

The piece shown is a wooden stand on which resides a metal sculpture. The sculpture is of a large sword, the blade of which transitions into a plowshare. An inscription on the blade reads "and they shall beat their swords into plowshares." This piece has been shown in many churches to the pleasure of as many as half a million people.

Dovetails

The dovetail is the classic box and drawer joint. Dovetails are a mechanical joint, meaning that they hold together mechanically without glue or other fasteners. However it's best to glue these joints because they will come apart if banged the wrong direction. Use this joint where two broad pieces of wood join at right angles with the grain direction perpendicular to the joint, not parallel to it.

Miter Joint

A picture frame joint is a miter joint. Two pieces cut at an angle (usually 45 degrees) are joined along the angle. You can join these in different ways, such as mortises and tenons, dowels, splines, or

"The motivation to create beauty for me lies in the fact that we become the results of what we do, so that my efforts to create beauty end up yielding aesthetic pleasure. There's a certain joy to it, and if you go far enough with it you become competent, and competency is very rewarding. . . .  What I wanted was happiness, beauty, and truth around me. I'm very happy."

even nails. Ultimately, miter joints are less strong than joints that don't involve angles, but a tight-fitting miter looks nice.

Mechanical Fasteners

Nails—is there anything they can't do? It's so easy to just grab one and start banging! Nails are very useful in certain aspects of woodworking, such as attaching solid wood to plywood and vice versa. Small finish nails are ideal for attaching pieces of molding. Although purists scoff at nails in furniture, some of the finest antiques have handmade square nails gracing unseen portions of their construction. The main problem with nails is that they split wood, but you can prebore for nails just like you do for screws.

Screws are useful in many situations where wood joins wood (see figure 18). But to use them correctly, you must bore proper holes for them. If you don't, the wood might split or the screw shear off as you try to drive it in. In addition, a screw in a proper hole is stronger than one with no hole. This is because the screw's threads need to contact solid wood, not a distorted surface split by the shank as it is blasted in. Get tapered drill bits with countersinks and make a wider, deeper hole for dense woods and a less wide, less deep hole for softer woods.

Liquid Joinery

Although many wooden-boat builders don't want to admit it, much wood joinery today is done with epoxy. Other wood glues don't make a strong bond if there is a wide gap between the pieces being glued, but epoxy does and thus is very useful in situations such as boatbuilding, in which wood pieces join at odd angles. Epoxy is not a good alternative for proper wood joinery in standard furniture construction, but in nonstandard situations it can work wonders.

Old Shoelaces

Make sure that any old shoelaces you use for wood joinery are not terribly worn. A good friend of mine once warned me, "Jeff, never, ever, run out of baling wire" as he repaired a chair on which I would sit during that night's poker game. My grandfather was good with stucco plaster and pitch, but that was before the days of auto body filler. The moral to the story is when all else fails, improvise.

Hand-Tool Techniques

Machine tools simply do what hand tools did, only faster and with the tool configured differently. The last point is important because

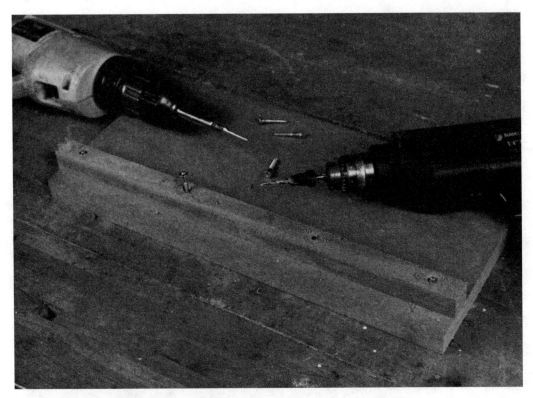

Figure 18. Installing screws is made very easy by first pre-drilling for the screw with a tapered bit and countersink, followed by sinking the screw with a hand drill that has a variable speed trigger.

you'll sometimes find that the configuration of a machine tool does-n't let you do what you need to on a piece of wood, but that a hand tool might. For example, it's generally unsafe or impossible to run very short parts through a table saw, jointer, or planer. When you need to make short parts with these tools, it's best to make long ones and cut them short afterward. But what if your parts are short to begin with? You can safely work them with hand tools. How do you smooth a board that's too wide for your planer? You can use a belt sander or a hand plane. How do you make stopped cuts on the inside of a piece that's too wide for the throat of your band saw? Use a handsaw. Hand-tool skills will help you a lot in the long run and are well worth the effort involved in learning them.

The key to hand-tool skills is keeping the tools sharpened and tuned correctly so they work. A band saw with a less-than-sharp blade that isn't tracking quite right can still be used well enough, but a dull hand plane that isn't tuned up is worthless. My favorite hand plane was in a sorry state when I got it used at a yard sale, but I used it to learn how to tune up a plane, and since then I have done a lot of fine smoothing with it. Let's look at how to tune up a hand

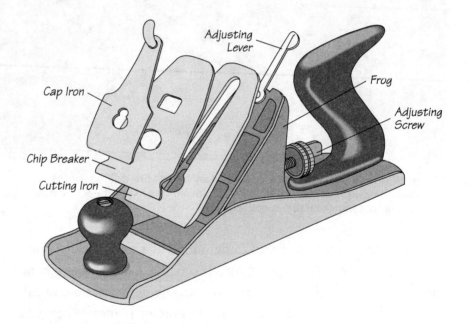

Figure 19.   Hand plane components.

plane (see figure 19). Shape the sharpened edge of a plane iron to a very slight arc so that the middle dips into the wood before the edges. When the corner of a sharp plane iron dips into the wood, it leaves a little ridge, so you want the iron to curve up and away from the wood at the edges.

If the sole of the plane is not flat, it can't work correctly. To find out how flat the sole is, get a piece of window glass and apply machinist's bluing to it. Remove the plane iron from the body, clean the sole with steel wool, and put the sole on the glass. Push it across the glass, then lift and see where it picked up the bluing. These are the high spots. If it's obvious that the sole is not very close to flat, put emery cloth onto a flat surface, such as a cast-iron machine table, and grind the plane sole on the emery cloth until the high spots are leveled.

On the plane iron, you'll find a chip breaker, which contacts the iron very close to the sharp edge. This piece of steel causes the shavings to curl away from the point of cut and reduces splitting ahead of the cut, reducing tear-out. Tear-out consists of unsightly indentations that are left in the surface because the wood was torn up ahead of the iron rather than shorn cleanly by it. The chip breaker must contact the iron closely along its entire width, or chips will get wedged into it, fouling the cut. Flatten the face of the chip breaker that contacts the iron on your stones so that only the leading edge of it contacts the iron. Set the breaker on the iron so that it's very close to the edge for fine work and as much as 1/16 inch away for rough work.

The frog is the little casting that the iron sits on. It's adjustable on all metal planes except block planes. Adjust the location of the frog so that the cutting tip of the iron comes as close as possible to the leading edge of the throat in the plane sole. This helps reduce tear-out as well.

It can be a bit frustrating at first to get a hand plane to work well, but keep fussing with it until you do. You'll be very proud of yourself, along with having a time-saver that will pay for itself over and over.

Other useful hand-tool techniques for the modern woodworker include using a jointer plane to "shoot" the edges of boards straight.

Handy Hint

Use a belt sander to grind a bevel on chisels and plane irons. If you use a hand belt sander to do this, be sure it is securely positioned so it will hold steady while in use.

You need a long jointer plane to do this, as your smoothing plane isn't long enough. Getting the edges of long boards straight in preparation for gluing them together can be a challenge even with a jointer and a table saw, so using a jointer plane is not so outlandish.

There are many other types of planes, each with a specific purpose. A rabbeting plane cuts the bottom of a rabbet, or groove, to deepen or smooth it and is a useful addition to machine rabbeting techniques because often a machine-cut rabbet needs to be smoothed or adjusted. The work of molding planes has largely been replaced by router bits, but you might find a molding plane that cuts a shape that you can't find in a router bit.

Developing your dovetail saw and chisel skills will help you greatly in many situations. In many cases, your machines will take you only so far, and being able to clean up that last little bit with a sharp chisel and maybe a cut or two with a handsaw will keep you moving along.

The spokeshave is still the tool of choice for shaping curved pieces (see figure 20). I recently made a router jig to shape some large cabriole legs, and it worked fine for the major portion of them. But I still had to shape the back of them with a curved-sole spokeshave, and the spokeshave was just the thing for rounding over the sharp edges of the legs along the curve.

Machine-Tool Techniques

The key to working well with any machine is understanding what it was designed to do and working within its mechanical capabilities. The first thing I do when I see an unfamiliar machine is look at it until I understand how all of its components work. Be sure before you use any machine tool that you understand how the makers intended it to be used and that you have it set up correctly.

Figure 20. Use a spokeshave to smooth and shape a rounded surface.

Carbide Cutting Tools

Until recently, all saw blades and other cutters were made of steel. Steel can be sharpened to a razor edge, but it doesn't hold that edge very long. Steel cutters suffered badly as industry increased its use of manufactured panels, such as plywood and MDF (medium density fiberboard), which contain a high glue content. Glues wear down steel much faster than wood. Carbide cutting tools were developed to address this problem. Carbide is a composite material made from fine powders of very hard, high-carbon steel mixed with a binder that holds them together. Small pieces of carbide, called

Safety, First and Foremost

Safety should be the first thing on your mind at all times when doing woodworking. The most precious and important tools in your workshop are your fingers and your eyes, and each of your tasks should be organized with safety in mind. Experience has shown that accidents happen "at that one moment I wasn't looking." Although it's always easy to find a way to do a woodworking procedure safely, it's also easy to find a way to hurt yourself through lack of attention to what you are doing.

Although it's difficult to generalize about safety procedures because each machine and each separate machine setup is different, the main things to think about are stabilizing the work and the proximity of your fingers to the cutting action. On all setups, be sure that you have a way of holding the work steady while it is being cut so that it does not come loose and get thrown by the cutter. Otherwise, if your work does come loose, your hands might have nowhere to go but in harm's way. Always keep your fingers as far away from the cutting action as you can. This often involves using hold-downs or push sticks that allow you to push the wood from a distance. You should always have eye, ear, and breathing protection available to use whenever you need it. One of my main rules of thumb is to always keep my eye on the cutter as I do the operation. Doing this ensures that my basic instincts won't let me put my fingers too close to the work.

Now a word about fear of the machines. Fear is a very healthy thing in all of us, as it protects us from danger. At the same time, if you're so afraid of a certain operation that you can't do it in a stable manner, you might be creating a dangerous situation. Confidence in your ability is a key ingredient to safely and effectively operating a machine. If you don't feel confident with a certain operation, work up to it with simpler tasks until you know that you have the feel of it and can do it safely.

tips, are brazed onto steel bodies of saw blades and router bits to make the tool. Carbide stays sharp far longer than steel because the fine granules are so hard. Ultimately, steel can be brought to a sharper edge than carbide because steel is a uniform material, not a composite, but finely ground carbide is sharp enough and holds the edge so long that it's worth the extra expense.

When you shop for carbide tools, the main issue to pay attention to is the quality of the grinding that was done on the carbide tips. A million other factors determine the ultimate quality of carbide tooling, but these factors are relevant only to large factories that beat their carbide tools to death on a regular basis. Many hobbyists might never need to sharpen their carbide saw blades. All you need is a blade that was sharpened well to begin with.

The quality of grinding is entirely a function of how smooth the manufacturer made the two surfaces on the carbide tips that come together to make the edge. Both surfaces must be very smooth for the edge to be sharp. If one or both are ground roughly, the edge looks serrated under a microscope. Such a serrated edge doesn't cut as well and will become dull faster with wear as it loses relatively large chunks along the uneven edge.

You'll find that on most saw blades the grinding is fine because it's technically easy to fine-grind saw blade tips. The place to really look out for rough grinding is on curved router bit profiles. It's technically easy to grind smooth the flat face of the carbide tips in such a bit, but it's harder and more time consuming to get the shaped edge ground smooth. This is where some manufacturers cut corners, and this translates into little nicks in your wood surface and a bit that will need sharpening more frequently (if you use it enough to get it sharpened at all).

Other things to look for in carbide tools are voids in the braze and bad balancing. Any holes in the brazing that holds the tip to the body that are larger than a pinhole could mean that the carbide is not properly attached to the body, and this is potentially dangerous. If you notice that your router bit or saw blade is out of balance and causes excessive vibration, do not use the tool and return it to the retailer.

Handy Hint

An unscientific but simple and reliable test for the quality of carbide grinding is to run a pencil tip along the ground surfaces. On a finely ground surface, the pencil tip will slide along without wearing down much at all. On a roughly ground surface, the tip will grind along like you were rubbing it across fine sandpaper. Avoid carbide that feels like this.

Table Saw

A wide variety of saw blades are available for table saws, but it's easy to narrow down the choice to only a few that you'll need. I keep a steel blade around for cutting dirty wood because I don't want to mess up my good carbide blades and steel is cheap to sharpen. You can get different carbide blades for ripping (cutting with the grain) and crosscutting (across the grain), but I get by with a carbide combination blade that does both adequately. If you cut a lot of laminates or plywood, you might look into special blades that reduce the tearing that occurs on the edge of the cut with these materials.

A dado cutter (see figure 21) makes a dado (a groove in the middle of a board) or a rabbet (a groove in the edge of a board) that is wider than your other blades. They are adjustable to cut dadoes from ¼ to 13⁄16 inch wide. Good carbide dado sets cost a bundle but are worth it if you plan to use them a lot, say, for making lots of plywood cabinets. I get by with an old steel dado set that I got used from my tool sharpening man. I use it rarely and haven't yet had to get it sharpened.

Use the fence on your table saw to accurately rip boards to the width you need them to be. This fence controls the distance from the blade to itself and can also be used to locate a dado cut in the middle of a board. Sometimes you'll want the blade to be right up against the fence, as in cutting a rabbet on the edge of a piece. In this case, attach a wood cover to your fence so the blade cuts that and not the fence itself.

When using your table saw with the rip fence, always be sure to keep the wood down on the table so that it does not come up and get on top of the blade. Begin the cut by pushing the work against the fence on your side of the blade, then simultaneously pushing into the blade. Pressure is thus applied at a 45-degree angle, keeping the wood against the fence and moving ahead. With one hand keep

Figure 21. A dado cutter is a set of matched saw blades that you stack together on the table saw spindle for making wide cuts, such as dadoes and rabbets.

pressure on the side of the piece so that it stays against the fence, and with the other hand push it ahead. As you finish the cut, remove the hand that is applying the pressure to the side and push the work clear through with the other hand. On pieces less than 2 inches wide, use a push stick to complete the cut so that your fingers don't get too close to the blade.

To cut across boards rather than along their length, remove the fence and put on a miter gauge (see figure 22), which fits into a slot in the tabletop. Set the gauge at 90 degrees for square cuts or at any angle up to 45 degrees. To cut beyond 45 degrees, you'll need to make your own miter jig out of a piece of plywood with a runner on the bottom that fits into the miter gauge slot. Screw fences to the top of the plywood to hold the work in place as it is cut.

A more accurate way to do square cutoff and angle cuts on the table saw is with a cutoff box (see figure 23). This is a larger, more elaborate variation on the miter jig described previously. Start with a piece of plywood at least 2 feet square and put runners on the bottom of it for both the miter slots on the table of your saw. Then attach 2-by-4s across the front and the back, on edge and at 90 degrees to the travel of the blade. Hold your work on the 2-by-4 closest to you and push it through the blade, holding the piece steadily against the blade. The larger jig will give you more accuracy and versatility for making a variety of angle cuts. But the problem with a cutoff box

Figure 22.   Use your miter gauge on the table saw to cut angles on the ends of pieces as shown.

Figure 23. Build a cutoff box for your table saw for making accurate cut on the ends of pieces.

is that it's hard to handle pieces much longer than 3 or 4 feet. For that you are better off with a radial arm saw.

You can cut tapers in boards on a table saw with a jig, such as the miter jig described earlier. The jig, which can run along the rip fence, simply holds your work at a slight angle to the blade along its length. Be careful with a jig like this so that you don't cut off the bottom of the jig as you make the cut, and be sure that your fingers are not in the path of the blade.

A variety of special setups are possible on a table saw. A tenoning jig holds pieces on end, sticking straight up vertically, so that the blade can cut tenon faces on the end. These jigs slide into the miter gauge slot. It's possible to make most of the cuts needed for dovetails with the miter gauge. Special molding cutter heads that are made for table saws have interchangeable steel cutting tips, allowing you to make shaped profiles. But these are potentially very dangerous, and they do little that you can't do with a router.

Band Saw

Most band saw blades are made of steel, but more expensive types, called bimetal, have harder steel on the tips than on the band. Some woodworkers swear by these; I've always been satisfied with the regular steel ones, which cost less and are disposable. Use blades of lesser width to make cuts with tight curves (¼ or ⅜ inch wide) and use wider blades to make straight cuts or to resaw thick boards into thin ones (½ or ¾ inch). Blades with fewer teeth per inch (three or four is the minimum for smaller saws) cut faster because they have more chip clearance, but they leave a rougher surface. Blades with many teeth per inch can't cut as fast, especially on thicker woods, because the chips get clogged in the small space between the teeth before the blade is out of the wood. But these blades leave a smoother surface and are best used for joinery setups where accuracy is important.

The cut on a band saw must always be into the blade, not at an angle to it. When cutting curves, turn the work as it's being cut, not before you push it in. If you twist the blade, you can cause it to come off its guides or break it. When you have the table tilted, be careful to keep the work going directly into the blade and not falling sideways against its side.

You'll need to let your fingers come fairly close to the blade when cutting smaller parts. Keep your eyes on the point at which the blade hits the wood so that you can always see where your fingers are in relation to the blade. Once you begin a cut on a band saw, you should not retract the work with the saw running because you can easily pull the blade off the wheels. If you need to retract the cut, turn off the saw and wait until it stops. You'll find that with woods ¾ inch thick or less, you can retract the work with the saw running if the cut is not along a sharp curve and you retract slowly, allowing the blade to track on its wheels and stay within the guides.

Cutting out cabriole legs on a band saw is so much fun that you should cut one out even if you aren't going to use it (see figure 24). Get a thick, square piece of wood and trace the shape of the leg onto two sides of it. It's best to use a template to do this so that both sides look the same. Now cut out one of the two sides. You'll notice that you've just cut off the side that has the other tracing! No problem—just tape it back on with masking tape and cut again. Now you get to smooth the leg with a spokeshave or a stationary sander. A rasp comes in handy for tight spots. To finish it, shape and smooth it with a scraper or successive grits of sandpaper.

Did you know???

Cast iron tools can warp. For a few weeks or months after manufacturers pour the cast iron for your table saw or band saw table, the iron moves a bit, until it has cooled solid. The best manufacturers let the castings sit a while to work out these stresses before they mill them flat.

Resawing on a band saw is a bit tricky, but with practice you'll get the hang of it (see figure 25). Your object is to split boards along their thickness accurately, making thinner boards. Before you begin, look carefully at the stock that you plan to resaw. If it's not straight, the resulting boards won't be either. Garbage in, garbage out, the saying goes. Perhaps you don't need the boards to be very straight, but remember that the resulting boards might be even less straight after resawing because you can release tensions in the wood by splitting it in two, causing the pieces to move.

It's best when resawing to use a jointer and a planer in conjunction with the band saw. Face joint the stock so that it's flat, then plane the other side so that the pieces are of uniform thickness.

Figure 24.   Cutting a cabriole leg on your band saw is one of the most fun things you'll do with the saw.

Now resaw, placing a flat face on the fence. After resawing, you might want to face joint again to flatten the pieces a bit before planing them.

If you don't have a jointer and planer, you must pick straight stock to begin with or be willing to live with the result. Somehow you'll need to clean up the rough surface left by the band saw blade. A smoothing plane is a good choice, or you can use a belt sander or get in good with that guy down the street who just got a planer.

Use a wide blade with few teeth per inch for resawing, such as a ¾-inch-wide blade with three or four teeth per inch. Mount it in the saw, adjust your guides, and be sure to set the table at 90 degrees

Figure 25.   Use the band saw to split boards down their thickness, which is called resawing.

to the blade. Make a fence that holds a board or piece of ply-wood vertically and parallel to the blade. The width of this fence (its height off the table) is ideally just less than the width of the boards you're going to resaw. Clamp the fence so that it's parallel to the path of cut taken on the saw, then cut a test piece. Does the blade wander to the right or left? If it does, angle the fence to compensate. The fence must be dead parallel to the direction that the blade wants the work to come at, and testing is the only way to tell what that direction is.

Once you're set up, cut slowly. Allow the blade to cut at the rate it wants to so that you are not applying excessive pressure onto it.

This excessive pressure can cause the blade to deflect, ruining your cut. Hold the work steadily against the fence as it goes through the cut. With a sharp blade and a good setup, you'll be very pleased to watch as you split your boards accurately into thin ones.

You can cut tenons on the band saw with a few simple setups. First, set up a fence like the resaw fence described previously or like a table saw rip fence. But clamp a stop block on the other side to limit how far you can push the wood. Use this setup to cut the faces of the tenons. Then remove all this and set up a miter gauge for doing crosscutting. Again clamp a stop block to the table, but this time to the side of the blade. Use such a setup to cut the shoulders of the tenons. Note that the first setup is the one that determines the thickness of the tenons. You must carefully adjust the distance from the fence to the blade to adjust the tenon thickness. But to check the tenon thickness, you need to make the shoulder cuts to get rid of the waste. You don't want to break down the tenon setup to do the shoulder cut, so set up the shoulder cut on the table saw. This way you can make numerous test tenons without breaking any setups and then cut the actual parts.

Many other special setups are possible on a band saw. Methods for cutting accurate dovetails with a band saw have been developed. You can cut out precise circles by pivoting work on an axle of some sort. Because the band saw can cut very thick work, you might need to do miter cuts on a band saw that are thicker than your table saw can handle. You can put very thin blades on a band saw to do scroll work, but those 3/16- or even 1/8-inch-wide blades are so prone to breaking that they can be more frustrating than effective.

Radial Arm Saw

Like the table saw, it's nice to have a cheap steel blade to use for those occasions when you want to cut up some raspy lumber, but most of the time I keep a carbide crosscut blade on my radial arm saw. Putting a dado cutter onto one of these tools lets you easily cut dadoes across any boards you put on its table and is handy for doing shelf joinery.

Setting a radial arm saw so that it cuts accurately can be tricky because it has several different adjustments, all of which affect its accuracy. The blade must be parallel to the line of travel that it takes, and the blade must be perpendicular to the table. Your saw will have Allen screws or knobs that allow you to make these adjustments.

Accurately setting up the extension tables that you locate on one or both sides of the saw will help make it fast and easy to get a good crosscut. Carefully make the table flat, then set the fence perpendicular to the blade travel. Securely screw it all down to the floor and/or wall so that it stays in place.

One of the easiest ways to cut out a number of pieces all at the same length is to clamp a stop block onto your extension fence at the required distance from the blade (see figure 26). But before you cut off each piece to length, you need to cut the first end off accurately; however, the cutoff block will be in your way because the pieces are still longer than that. The solution is to place a wide scrap chunk against the fence at the point of cut and then to put your work against this piece. This scrap piece is wide enough to hold your work away from the clamped stop block so that it isn't in the way for the first cut. Then remove the scrap piece, flip your work piece (placing the freshly cut end against the stop block), and cut it to length.

Oops! You located the stop block a little too far away, and the piece is cut off too long. Loosen the clamp just a bit, but not all the

Handy Hint

The main safety issue to be aware of when using a radial arm saw is to never put one of your hands or arms across the blade's path of travel. Your left hand goes on the left side, holding the work against the fence, and your right hand goes on the saw handle, pulling the tool into the work. Pull steadily and evenly, always pushing the motor and blade back into the starting position when the cut is finished.

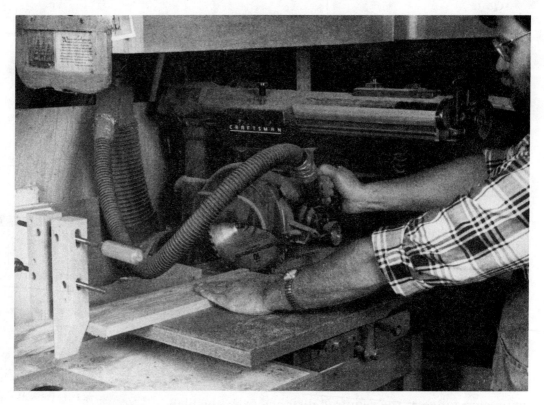

Figure 26.   When you need to cut a number of pieces to the same length on your radial arm saw, clamp a stop block onto the fence and use that as a reference on all the cuts.

way. Very gently rap the end of the stop block with a hammer to move it just slightly. Tighten up the clamp, then cut again.

What do you do if you cut the boards too short? My cousin has been trying to design a board-stretching machine that he claims he can patent and get rich with. But his research has been delayed by "technical problems," and we aren't expecting the prototype to be available for testing soon. Always cut your boards too long at first, then adjust your setup little by little, making test cuts as you approach the final dimension.

Router

A large part of the versatility of a router is due to the wide variety of available bit types and shapes. Bits are either straight fluted or shaped. Router bit tips are also called flutes. Straight-fluted bits make a straight cut, with no curves in it, and are used for cutting rabbets, dadoes, other grooves, or straight edges. Shaped bits come in many kinds of profiles for a wide range of tasks, from rounding over an edge to making fancy moldings, raising panels, and making tongue-and-groove joinery.

You'll find both carbide and steel router bits offered for sale, with carbide bits being at least twice the cost of steel bits. Carbide bits are well worth the price if you plan to use the bit often because they hold an edge 20 times as long as steel. But you'll spend a lot of money if you buy all carbide bits, so consider steel if you don't think you'll use the bit a lot. It's possible to sharpen steel bits if you can get the flat face onto a stone. Just sharpen on the flat face, but don't try to hone the curved edge. You can't sharpen carbide bits because they're too hard. Remember when you buy carbide bits that the quality of the edge grind is critical for the quality of the cut and the life of the edge.

Let's take a look at some basic procedures for different types of router use.

To round over the edge of a board and shape its edges, use a bearing-guided bit mounted in the router with the machine held by hand. Firmly grip the handles of the tool and be sure to hold the bit away from the wood when you start the machine. Then ease the bit onto the wood, allowing the bearing to rub against the wood but without applying hard pressure onto it. Keep the router moving one direction or the other; don't let it sit idle in one spot, or it will burn, leaving a mark on your wood and possibly causing the bit to wear faster because of the excess heat.

You can also shape edges in a router table. In this case, you push the wood into the bit rather than the bit into the wood. Keep your fingers away from the bit. If you must do small parts on a router table, hold them with something else, such as a wood thumb-screw, so that you can keep your fingers at a safe distance. Climbing the cut on a router table is difficult and should be avoided if possible. When you must, apply downward pressure to hold the work firmly against the table as you cut and go slowly.

Router Joinery

Many different types of joinery can be done with a router held by hand or in a router table. Mortise-and-tenon joinery lends itself well to router work because mortises are fairly easy to cut with a plunge router and tenons can be made in a number of ways with or without a router.

A plunge router has its motor mounted in the base on springs and telescoping posts so that you can push the bit directly down into the work (see figure 27). To use this capability to cut mortises, make a jig that holds the router centered on the edge of the piece to be mortised. The jig must allow the router to slide up and down the length of the piece just far enough to make the length of the mortise.

You can cut tenons with a router table and a special jig that holds the parts to be tenoned vertically with one end on the table. This is similar to the tenoning jig described for the table saw, which slides in a miter gauge slot. Use a straight-flute bit in the router table projecting no more than 1½ inches above the table. (Note that this height limits the length of your tenons.) The jig must be adjustable so that you can carefully regulate how far into the edge of the part the bit is cutting, and it must slide freely past the bit either in a slot or between two fences.

What Direction Should You Push the Router?

When you hold your router against the wood, you can push it to the right or to the left. Which is better? Well, it's a trade-off. Pushing it to the right is a "normal" cut, where you are causing the spinning bit to throw its chips ahead of the cut. Pushing to the left is a "climbing" cut, where you cause the bit to throw its chips behind the cut. This is called a climbing cut because the bit wants to climb onto the wood; if you don't hold it steady, it will grab the wood like a knobby tire with sharp teeth and take off. This can be dangerous, so when climbing the cut you always want to hold the router steadily and move it slowly, but not slowly enough to cause burning. In addition, don't do climbing cuts with large bits; use only bits with a ½-inch or smaller depth of cut.

The advantage to a normal cut direction is that you can cut faster without the possibility of the router climbing out of control. The main disadvantage is that the bit is more likely to cause tear-out when cutting this direction, especially when cutting against the grain. The advantage to climbing the cut is that it produces less tear-out, but it's harder to control and ultimately will cause the bit to lose sharpness faster, but only if you do a lot of this.

Special stile and rail bit sets are made for making kitchen cabinet doors, but they can be used for making frames for cabinet sides or other situations as well. These bit sets cut the pretty shaped edge that goes around the inside of the door or frame, and they cut the matching joinery profile at the ends of the rails (horizontal frame components), where they must fit over the shaped edge on the stiles (vertical components). Use these bit sets in a router table. You must do test cuts with these bits to adjust the fit of the stile and rail cuts to get a good joint.

Dovetail Jigs

As many as a dozen different dovetail jigs are on the market that you can use with a router to cut dovetail joints. The main differences

Figure 27. This special jig with a plunge router lets you make accurate, well aligned plunge cuts in boards for mortises.

between these different jigs are versatility and cost. The ones that will cut only one kind of joint are generally inexpensive, whereas those that cut very many are, you guessed it, much more costly. Before investing a lot of money in one of the versatile expensive jigs, think long and hard about how much you really plan to use it. They are well worth it if you plan to use them at all, but what good is it collecting dust on your shelf?

The inexpensive jigs make one joint: the basic dovetail drawer joint. These jigs are easy to set up and adjust. All dovetails are the same size and have uniform spacing. These are adequate if you have only one or two desks to make drawers for.

The Keller jig is an expensive aluminum jig that makes large dovetails for doing carcass work. You can get special bits with the jig that are high quality and that perform well. This is the best jig for joining large pieces to make large boxes, such as chests, desks, and so on.

The Leigh jig is an expensive jig with great versatility. It can cut perhaps the widest variety of joint types and configurations. You can vary the spacing of the dovetails in an infinite variety of ways and make other joint types. It takes some learning to get it set up, but you'll quickly get the hang of it. This is probably the ultimate tinkerer's dovetail jig.

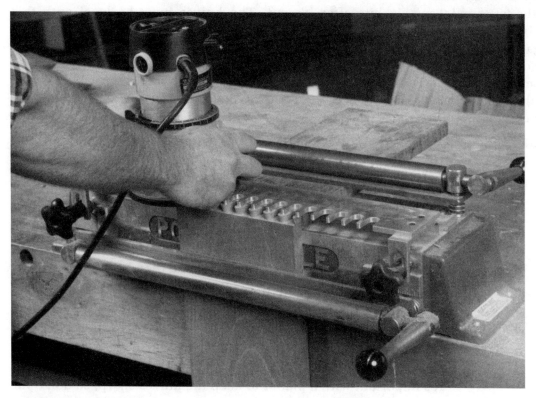

Figure 28.   Shown here is the Porter cable omnijig, a versatile and very rugged dovetail jig.

The Porter cable omnijig (see figure 28) rivals the Leigh jig for versatility, but its main advantage is ruggedness. This jig will also make a wide variety of dovetail joints and allow you to variably space dovetails. But its biggest advantage is as a long-lasting production tool. It will stand up to repeated use better than all the other jigs, except perhaps the Keller.

Handy Hint

When using a jointer, never pass your hands right over the cutter head. As each hand reaches the cutter, lift it up above the work and the cutter and then down onto the work again on the other side of the cutter. This simple procedure should become a habit when using a jointer.

Jointer

For a jointer to do its main job—cut a straight line—it must be adjusted correctly. The critical issue here is the height of the cutter knives in relation to the off-feed table, which is the table on the other side of the knives that the work hits after it's cut. The knives must be level with this table—not higher or lower. Thousandths of an inch matter here. That's why I suggested earlier that you get a jointer with an adjustable out-feed table because this makes it easier to adjust this setting. If your out-feed table is not adjustable, you must set the knives in the cutter head at exactly the right height to match the table, a difficult but possible task.

Along with straightening the edges and faces of boards, you can cut rabbets with a jointer by placing the fence near the close end of the cutters and dropping the in-feed table down to the depth of rabbet you need on a piece of wood. Cut tapers on table legs by placing the top of the leg on the out-feed table and then lowering the other end down until it hits the in-feed table. Then push it through. Do this multiple times to increase the taper (see figure 29).

Planer

It's important with a planer to keep the knives sharp, so don't put any dirty wood into the machine unless you plan to change the knives soon. Also be sure to keep nails and staples out of it. When you change the knives in a planer, the most important thing is to ensure that they are all set in the cutter head at the same distance from the cutter head's center. Most planers come with tools to help you do this. This applies as well to a jointer, the cutter head of which is similar to that of a planer.

Most of the work you do with a planer will be simply to reduce the thickness of boards to a uniform thickness, but with jigs you can make it do more. Taper table legs with a planer by making a jig that holds the leg at a bevel as it passes through the machine. Such a jig

Figure 29. You can use a jointer to cut a taper into a board by starting the cut in the middle of the piece.

is as long and wider than the leg itself, supports it well in front (in the back and in the middle), and passes through the machine with the leg set within it. Make a taper along the width of boards by making a fixed jig on the bottom plate of the planer that simply holds the board tilted as it goes through. Be careful with all such jigs that you don't take too much of a cut in one pass and that you don't cause the cutters to try to cut a nail or screw in the jig.

Drill Press

A drill press might seem like a safe machine, but it can be dangerous. Remove jewelry and watches before using it (as you should before using any machine) so that they don't get caught in it. Never bore into any wood or metal without the work being secured on the drill press table with a fence alongside it or with the work firmly clamped down.

Your drill press has adjustable speeds that you change by moving the belts up or down on the tapered pulleys above. Use the highest speed (2,500 revolutions per minute, or rpm) only with your smallest bits (1/16- to 1/8-inch diameter). Use the slower speeds (500 to 1,000 rpm) for bits 1/2 inch or larger. Don't use very long, skinny bits in a drill press, as they can bend and whip around like a wire. Use these with a hand drill.

A chisel mortising attachment for the drill press is a hollow square chisel that mounts onto the drill press collar above the chuck (see figure 30). It has a round bit inside the hollow square chisel. The bit spins and clears the waste while the chisel corners punch out the corners of the square hole. The chisel portion of the tool must be very sharp for this to work well. Use a round, tapered grinding stone to sharpen the inside of the chisel. You can find these with shanks that allow you to mount them to a hand drill. Use a chisel mortising attachment to cut mortises in the edges of frame parts for

Figure 30. Shown here is a chisel mortising attachment for a drill press, which allows you to use the machine to cut square holes for mortises.

tenons or for any other square holes, such as mortises in the face of a wide box side for shelves with corresponding tenons.

Small drum sanders with diameters of ½ to 1½ inches can easily be mounted in a drill press for use in sanding curved edges. These little rubber cylinders have metal shanks in the middle and replaceable sandpaper cylinders that fit around the rubber. Although you can get finer grits of sandpaper for these, they still tend to leave a rougher surface that needs hand sanding or scraping. Use a drum sander mounted on the drill to fair in an uneven curved surface after the band saw, then finish smoothing the surface.

Lathe

When you're learning to use a lathe, use a face mask. The potential danger with using a lathe is that a chunk of wood will fly off the spinning work and hit you. This can occur when you use the tool incorrectly and dig it into the work as it spins, causing a chunk to fly.

It can also occur when the wood is cracked or improperly glued. Until you have the experience to know that your technique is safe and that the wood is not going to come apart, play it safe and use a face mask.

My first project on the lathe consisted of four table legs for a simple pine table. Pine turns nicely, even the knots, and it looks pretty. I bought 2x (two-by) knotted pine and glued together two pieces per leg, so each started as a 4-by-4. I used a gouge to rough out each cylinder at the lathe's slowest speed, then sped it up a bit to cut the details and make it smooth. This fun project gave me the confidence to move on to other projects on the lathe.

The trick with turning is how you hold your tool on the work. It should contact the work in two places: the cutting edge and the back bevel. The back bevel rides on the surface that has just been cut by the cutting edge and can give it a bit of a burnish for a smooth finish. Riding the back bevel against the wood helps avoid digging the cutting edge into the work too far, causing the tool to grab and gouge out an ugly shattered chunk.

Handy Hint

One problem with small drum sanders is that the grit tends to clog up pretty easily. Special rubber sticks are made for cleaning this and other power driven sanding grits. These sticks are a bit costly but last a long time. However, a workable alternative is to use a chunk of old rubber tire, believe it or not. Push the cleaner stick or tire chunk into the sandpaper as it spins to clean it.

But before you learn to use the tool this way, practice with a scraping action. Hold the tool on the tool rest at a 90-degree angle to the work and pointing to its center. Gently move the tool in and across to lightly scrape the surface. Always use this technique when roughing a fresh piece into a cylindrical form. Then use the bevel riding technique once the piece is round. First try the bevel riding technique with a gouge, which is easiest, and then try it with the more challenging skew.

Turning table legs as described here is done "between centers," that is, with the work spinning between two supports on either end. To turn a bowl, support the work on one end only (the headstock) with a faceplate or special lathe chuck (see figure 31). Now you can use your tools to cut out the inside of the bowl. To do this, you need large pieces of wood, which you can sometimes buy, but the prices will scare you. In addition, you can glue up large pieces from smaller ones or look for freshly cut trees in your area. Green wood turns nicely, but then it must dry; thus, turn out a green piece to the general shape you want, then put it above the kitchen stove for a few weeks. Once it has mostly dried and changed its shape a bit, finish turning it to the final form.

Sanders

I promise to say it only one more time: You are better off learning to use a smoothing plane and scraper for finish work than sanders because these hand tools do the job in less time. But because many people are intimidated by hand tools, power sanders are here to stay.

Belt sanders can remove a lot of material quickly. Carefully adjust the tracking of your belt on the tool before you set it on the wood and observe the tracking of the belt as it's working. The tool will have a small knob to adjust the tracking, which can get off balance quickly. Hold the sander evenly on the wood surface so that it

Figure 31. This special chuck mounted on a lathe lets you grab your work in its jaws from one end only, allowing you to work the other end.

does not tilt one way or the other. Always keep the belt's movement parallel to the grain, or you'll see the ugly cross-grain scratches that are the hallmark of power sanding technology. Although you can get 220-grit belts for a belt sander, the tool is best used with 180 grit and lower and does it's best work with heavy grits that remove a lot of material quickly. To finish sand, move to an orbital sander.

Orbital sanders are smaller units that you can hold in one hand. They are not as good at removing large amounts of material with heavy grits but are far better at finish polishing with finer grits. Belt sand to 180 grit, then start with 120 or 180 grit with an orbital sander and work to at least 180 grit for the finish. Polish only to 320, 400, and 600 grit if the wood is highly figured. With burl, crotch, and other figured woods, such fine polishing gives the wood three-dimensional depth, but on ordinary wood the visual improvement achieved with such grits is minimal; 180 or 220 grit is enough for these woods.

A good way to get a lot of sanding done is to visit a local cabinet shop that has a wide belt sander. These $10,000 industrial machines make production work easier for the shops but cost so much that they need to rent them out to make them pay. Shops will charge you about a dollar a minute, and in half an hour or so you can bring a couple of dozen ugly boards to a beautiful finish. Here's where power sanding is faster than hand tools.

Finishes

You need to consider three main issues when deciding which finish to use on your woodworking: appearance, durability, and ease of application.

Wipe-on oil finishes give a natural appearance, which is to say that they don't build a thick layer on top of the wood and so don't look like a film covering the wood. They absorb into the top layer and make a very thin film on the top. This kind of finish is not very durable, so it is not appropriate for any tabletop that will get wet and be worn with use. It's easy to fix a marred oil finish by rubbing with steel wool and then applying more oil. Wipe-on oil is easy to use because you just wipe it on and wipe it off.

It can be applied very thin, it looks very good, and it is the least "plastic" of the film finishes. Shellac is very durable, and it seals the wood well against moisture variations in the air. But shellac has a

---

Handy Hint

With orbital sanders, watch for swirl marks, which are the orbital's hallmark. You can't grind out a 100-grit swirl mark with 220-grit paper unless you want to stand there for an hour. As you move through each successive grit, carefully check that you have removed the devastation left by the previous grit. This rule applies to all sanding, power or hand.

What's It Going to Cost?

If you went right out and bought one of each of the tools described in this chapter, you would go broke in a hurry! I certainly didn't buy all my tools and machines at once. I was a yard sale hound for a long time, and much of what I have I accumulated gradually, especially the small hand tools. Still, if you want to get started soon, you can't rely on what you might find by chance around the neighborhood, so you'll be looking to buy new or used from the want ads. What's it going to cost? Let's look at three scenarios: a beginner to the craft, someone with intermediate skills, and an advanced woodworker.

Beginner

Let's say that you have no tools at all and that you decide you want to build a few small boxes, such as that shown in the next chapter. You'll need a table saw, a bench, and a few hand tools. Because a table saw is a fundamental tool for any shop, it's best to get a good one from the start, so here's where your largest initial investment should go. You can get a new 10-inch saw with a 2- or 3-horsepower motor for $1,000 to $1,300. A used one will cost around $700 to $800. You can get a new lighter-duty contractor's saw for about $700 and a used one for around $500 or less.

Then you'll need a bench. You can build a simple bench with 2-by-4s and plywood for less than $100, or you can buy a bench through a woodworking catalog for as little as $200 or as much as $2,000. Then you should be prepared to spend a few hundred dollars on smaller items, such as chisels, saws, hand drills, and screwdrivers.

So you found a pretty good table saw in the want ads for $700, decided to get a $500 bench from a catalog, and bought a set of chisels and a block plane for $100. At less

major drawback in that it doesn't hold up to liquid. Thus, it's not the best for tabletops or anything that might have, for example, a glass of water placed on it. Shellac is tricky to apply because it's dissolved with alcohol, which evaporates quickly. The advantage to this is that you can brush or spray many coats in one day; the disadvantage is that misapplied shellac looks blotchy.

than $1,500, you're off to a good start, having enough to make some boxes and to have some fun as well.

Intermediate

The next step is another machine, say, a band saw. You decided to get a quality new one for $750. Then you found an old, beat-up radial arm saw in a yard sale for $100—a good deal because it still runs okay. A new router cost you $125, then you promptly spent $75 on bits (mostly good carbide bits that will last a while). For just over $1,000 more, you have greatly increased your shop's capabilities. Total expenditures now are still under $3,000, and this can be stretched out over a period of time because you had enough to get started before you bought these.

Advanced

It's been a couple of years now since you got started, and you have a number of projects under your belt. You've been feeling the limitations of your machines because you don't yet have a planer to do thicknessing. You also wish that you had a jointer to straighten stock with.

You bought a new small planer for $400, but you just weren't happy with it because it did not have enough power. So you sold it to someone for $300 and used that to help pay for a new, more powerful one at $1,000. You found a used 6-inch jointer in the newspaper for $400 in good condition. Finally, you realized that you really ought to have a dust collector, which cost $350 with the flex hose and some attachments. With a few more hand tools, you've spent another $2,000, for a total of about $5,000, to round out your shop with machines that are capable of handling most of your woodworking operations.

Lacquer was developed as a commercial wood finish. It's easy to apply by spraying, dries quickly so that you can do it all in a day, and holds up to water and alcohol well, unlike shellac. It has a somewhat plastic appearance and is best applied by spraying, but some slower-drying formulations are sold for brushing. Lacquer is a good choice for tabletops but is not as hard wearing as oil-based varnishes.

Originally, oil-based varnishes were made from tung or linseed oil mixed with thinners and dryers. As the oil dries, it forms a very hard coat. The appearance is somewhat plastic but not bad. These are the most durable of the traditional finishes but take longer to apply because they take overnight to dry, so applying numerous coats can take days or weeks. Use these finishes if you want maximum protection with good appearance and are willing to take the time.

> ## Did you know???
> Shellac is a traditional finish made from the cocoons of the lac insect, which lives in Asia.

Many other oil-based finishes on the market are not made from tung or linseed oil. Polyurethane varnishes are one example. They are very tough and best for hard-use situations, such as dining tables. They look like what they are—plastic—and must dry overnight.

Many water-based finishes are now on the market because paint and lacquer thinners are polluting and bad to breathe. At first these finishes were hard to apply smoothly and looked very plastic, but they have gradually improved. However, don't assume that because they're water based they're safe: Other ingredients used in these formulations can be harmful. Be sure to use gloves and good ventilation as you would in any finishing operation.

The first step in getting a good finish is to smooth the wood well with successive grits of sandpaper to about 180 or 220 grit or by using hand tools. The next step is to carefully apply the finish. The first coat will raise the grain, that is, cause any loose fibers to stand up. Let the first coat dry and then rub out the surface with fine steel wool. Clean the surface and apply more coats, sanding with 300 grit or rubbing with fine steel wool between coats if any roughness is apparent. Apply only enough coats to get a slight buildup of finish (except with wipe-on finishes), but on tabletops apply a few more for extra protection. Finally, rub out the finish with extra-fine steel wool and apply a coat of furniture paste wax to give it a shiny luster.

Closing Thoughts

How many of these techniques will you need to pursue your objectives? Looking at them all at once can be rather daunting, but remember that you don't need them all at once. Depending on what you do, you might need only a few of the tools and a few of the skills described here. I gradually accumulated some skill in all the major areas of woodworking by taking on different projects one at a time. It took a while, but I didn't have to know it all in order to make nice things from the beginning. Neither do you, so just focus on what's ahead of you and leave all the rest for later.

Creating Your Woodworking Projects

IN THIS CHAPTER, WE'LL LOOK at four projects that you can do with a minimum of tools and skills. Feel free to use these designs as they are shown here or to embellish them with your own design changes. The jewelry box, knickknack shelf, and bookcase projects lend themselves to production setups, so if you want to, you can make a number of the same items in one production run for efficiency. Although the stool project doesn't lend itself to production work like this, it does lend itself to using unique local woods or highly figured pieces from stock lumber. Using one-of-a-kind pieces of wood this way will make each stool different from the rest.

Use these projects to get you started with your tools and to develop new skills. Having built a few of these, you'll gain the confidence you need to proceed with other, more complex projects. Now let's make something!

 Jewelry Box

Basic Game Plan

Tools: A table saw, chisels, and a hand drill

Materials: Stock 1x (one-by) lumber (¾ inch thick) from the lumberyard, two hinges, and screws

Time: A couple of days, maximum

Parts Cut List: Jewelry Box (see figure 32)

2¾ x 3 x 6

2¾ x 3 x 11

2¾ x 7 x 12

Here's a beginner's project that will help familiarize you with your table saw. This is the only tool you'll need to make this box, although you could embellish it a bit by using a router to put a curved profile around the edges of the top and bottom where the bevel is. The splined miter joints at the corners of this box are easy to make with the table saw and will introduce you to making small parts safely with a machine tool.

Use 1x lumber for the sides, top, and bottom. You can use it at the thickness it comes from the lumberyard, about ¾ inch thick. I used mahogany, but you can use whatever you like. When you look for your lumber, choose a piece or pieces that are fairly straight. Look for something with a beautiful figure to the grain and put the nicest piece on top.

Begin by ripping your stock to the two widths you need: 3 and 7 inches as in step 1. Rip the parts to width before you cut them to length because it's safer to rip long parts on the table saw than short ones. You need a straight edge against the fence of the saw, and if

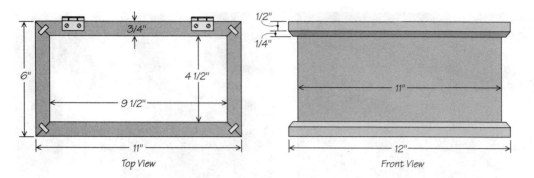

Figure 32. Cut list for Jewelry Box.

neither of the two edges are straight to begin with, place the straighter of the two against the fence and rip it over the width you need. Then flip the board and place the cut edge against the fence and rip again after moving the fence in ⅟16 inch or so. Do this a few times, and the edges will get straighter each time. Once you have a good edge, do your final rip.

Now set up your miter gauge as in step 2 to cut the 45-degree mitered ends of the four box sides. Make several test cuts with the setup to ensure that the miters are accurate. First check with a square that the miter gauge is set at 90 degrees so that the end is square, then check the angle of cut of the blade by doing the procedure shown in step 3. When two of the mitered ends are put together as shown, they should meet at 90 degrees, which you can check with the square. Adjust the angle of the blade until you get a good joint at 90 degrees.

Once you're happy with the setup, cut one end of all four pieces and leave the other end over length. Then clamp on a stop block on the miter gauge fence as shown in step 2. Set the stop block to cut the sides to the lengths you need, first at 6 inches for the short sides and then at 11 inches for the long sides. The stop block ensures that the corresponding sides will be exactly the same length.

Step 1. Rip stock to width for the sides while the board is still long, then cut the short parts out afterward.

Step 2. A miter joint uses angle cuts, such as here, where the blade is tilted at 45 degrees. The miter gauge is set at 90 degrees in this case, although it gets its name from its tilting the work at an angle to the blade.

Step 3. Test the adjustment of your blade angle by putting together a joint and laying a square to it as shown.

Step 4. Use this setup to cut the kerfs in the miter faces for the splines.

Now set up to cut the grooves in the miter faces for the splines that hold the sides together as in step 4. The stop block in that photo holds the parts in a given relation to the blade so that the cut will be made in the same location on all the miter faces. Don't raise the blade up too high, or you'll cut right through the piece! Leave about 3/16 inch from the groove to the outer surface of the wood.

Next, rip some stock to thickness for the splines as in step 5. Make a test cut or two to ensure that the thickness of the spline you make matches the thickness of the saw kerf in the mitered ends. When making thin cuts like this, use a push stick to complete the cut. Don't push your fingers right past the blade. Push the work through the blade the last 6 inches or so with a stick about 12 inches long. Let this stick get cut by the blade as it pushes the work through (better it get cut than you).

Make the width of this spline piece about 1¼ inches. Once you have this piece, you need to cut short pieces of it to length for the splines. Use a setup as in step 6 to do so. The most important element of this setup is the piece of wood clamped to the rip fence before the blade. Use this to adjust the length of each piece to be cut. If you were to simply put the rip fence at the distance from the blade equal to the length of spline you need, the blade would catch the cutoff piece between itself and the fence, throwing it dangerously. You need distance between the blade and fence after the cut is made, and the clamped-on piece allows this. After you cut off each piece, use a push stick to push the piece beyond the blade if the miter fence does not.

Make the length of each piece twice the depth of the kerf cut into the faces of the miters. Now you might be wondering why we didn't just rip the long piece to a width equal to twice the kerf depth, cut off pieces of this 3 inches long, and then use those as splines. If you did that, the grain would run parallel to the joint and the splines could easily break along the grain direction. The grain needs to run across the joint for strength, which is why we are cutting off short pieces across the grain as shown in step 6.

Glue up the sides with two splines in each joint as shown in step 7. Apply light pressure with the clamps, just enough to pull the joints together. Wipe off any excess glue with a rag and warm water.

Now use your miter gauge cutoff setup once again to cut to length the top and bottom at 7 by 12 inches. Now set your table saw

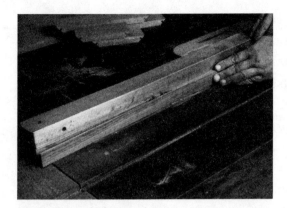

Step 5. Rip stock to the thickness required to make the splines as shown.

Step 6. Cut off splines from the ends of your thicknessed stock as shown. Pull the miter gauge back behind the scrap wood that is clamped to the fence, then butt the spline stock up to the scrap wood. Now, holding the spline stock against the miter fence, push through the blade.

Step 7. Apply glue to the splines and their kerfs, then assemble. Put one spline on the exposed end of each joint.

Step 8. Use the miter fence to guide the bevel cut on the ends of the top and bottom.

blade to 45 degrees and cut a bevel on the edges of the top and bottom along the long sides, then do the same on the short sides with the miter gauge guiding the cut as in step 8. Sand these beveled surfaces smooth.

Step 9. Chop shallow mortises for the hinge leaves using a sharp chisel.

Step 10. Use screws to attach the base to the sides as shown.

Step 11. The finished Jewelry Box.

Get a couple of small hinges and use them to scribe onto the wood the location of mortises for the hinges, then cut out those mortises as in step 9. Make these cuts deep enough that the top will contact the upper edges of the sides but not too deep, or it will bind. Pre-drill for the screws that hold the hinges, test-install the hinges, and then remove them for your finishing procedures.

To attach the bottom to the sides, clamp these together as in step 10 and drill for screws as shown. You need only one screw in each side. Countersink the screws so that they are recessed below the level of the wood and thus won't scratch any surface the box sits on.

Now give the box a careful finish sanding, gently rounding over all sharp edges. You might need 80-grit sandpaper to sand out bad saw marks, then use 120 and 180 grits to further polish the wood. Use a wipe-on oil finish, giving it a few coats over a few days to let each coat absorb and dry. Finally, reinstall the hinges and start thinking about who the lucky recipient of your fine efforts will be (see step 11 for finished jewelry box)!

 # Three-Peg Stool

Basic Game Plan

Tools: A band saw to rough shape the seats and a spokeshave to finish it; a belt sander and a hand drill; and a lathe (helpful but not necessary)

Materials: Stock at least 1¾ inches thick for both seat and legs

Time: About one day for one stool

Here's a fun project that gives you a chance to try out a few handtool skills where precision doesn't matter. You can make a unique stool, depending on your choice of wood. This one is made of California walnut. Use highly figured local woods that aren't available in the lumberyards or use knotted sections that would otherwise get tossed in the firewood box. You'll need a band saw to rough out the chunks as well as a drawknife and/or a spokeshave. A lathe helps but is unnecessary, and you'll need a belt sander to smooth out the lumber.

What? Earlier in the book, I said that there's little you can do with a belt sander that you can't do with a hand plane, and now I'm telling you to use a belt sander? When you're using highly figured

Step 1. Clamp a stop block behind the work on your bench as you belt sand it, or the belt sander will throw it backward. Start with 60- or 30-grit sandpaper if the wood is very rough, then use 80 or 120 grit to smooth it.

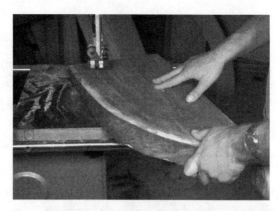

Step 2. Use a band saw to shape the seat of your stool. A ¼-inch-wide blade will allow you to cut tight curves along the edge if you want.

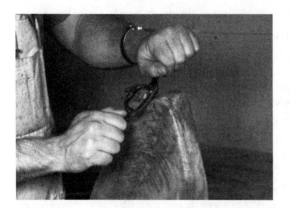

Step 3. When rounding over the edges with a spokeshave, be careful to follow the grain direction to avoid tearing out the surface. Whenever you use a spokeshave, sharpen it well first.

Step 4. Use a combination square to show you the angle at which to bore the holes. Before you start to bore, remove the square and hold the drill firmly with both hands.

chunks of wood or wood that has big knots in it, using a hand plane can be a real pain in the neck because the grain will tear a lot. So, yes, here is where a belt sander saves the day. Start the project by smoothing out your wood as in step 1 to see what it looks like.

You might be tempted to take the pieces to a planer to do this, but if they are knotted or highly figured, that's not such good idea. A planer will chew up such wood because of the variable grain direction, and such pieces are also likely to be not so flat, in which case they won't go through the planer well. Hooray for belt sanders (in certain situations).

Once you have cleaned up the wood well enough to see what the grain looks like, take it to the band saw and cut out the stool seat as in step 2. Cut away sapwood and any sections that are loose and structurally unsound. A large, loose knot in the center can work, as long as the surrounding wood is strong and you locate the legs in this sound wood. Cut out your seat in whatever shape you want or to a rough oval. Let the grain patterns in the wood show you what shape to make the seat.

Did you know???

Old post-and-beam barns were held together with hand-hewn pegs, similar to what you will make for this stool. The large beams were joined with equally large mortise-and-tenon joints, held with pegs that the makers cut with a drawknife. These pegs weren't perfectly round or smooth, but they fit the holes snugly, which is all that was required.

Now chase after the edges with a spokeshave as in step 3. Your goal is to round over the edges, but if you have large areas to remove that you didn't get to with the band saw, you can start with a drawknife (shown in photo 7). This tool is more rough than a spokeshave because it can pull off very large chunks at each stroke. A spokeshave can shave off only thin peelings, so it removes less wood but leaves a cleaner surface and is easier to control. Clean up the nastiness left by the drawknife with the spokeshave.

Now bore 1-inch-diameter holes for the legs as in step 4. Cut these at about a 15-degree angle. You can use a combination square as shown to align this angle, or you can simply gauge it by eye. Start the holes on the top of the seat because the wood will tear a little where the bit comes out, and you don't want that on the top. Use a spade bit as shown, or use a Forstner or auger bit.

Step 5. If you have a lathe, you can accurately turn the leg ends; otherwise, use a drawknife. On the lathe, use a calipers as shown to show you how close you are to the correct diameter.

Step 6. Cut slots in the leg ends on the table saw as shown. Set the blade to cut just off center so that after you have made two cuts, one on each side, the slot will be a bit wider than the blade thickness.

Step 7. When using a drawknife to shape the legs, stand back and be careful not to pull the tool onto your arms or body. Use smooth, long strokes rather than short, jerking ones.

Step 8. After gluing, use a fine-toothed dovetail saw to cut off the leg ends. Use a block plane, a sharp chisel, or even a belt sander to flush down the ends.

Now start on the legs. Get out three pieces at about 1½ by 1½ by 14 inches. You can turn the ends of the legs as shown in step 5 to bring them to the 1-inch diameter required to fit the seat holes. If you don't have a lathe, don't worry because the ends don't have to be perfectly accurate to make a strong stool. Shape the ends with a

Step 9. The finished Three Peg Stool.

drawknife or spokeshave until they just slide in the holes. They don't have to be perfectly round, just snug.

Next, before you round over the rest of the legs, cut slots in the ends on the table saw as in step 6. Raise the blade up as high as it will go so that the leading edge is as close to vertical as possible. Make two cuts, each slightly off center. Flip the part 180 degrees between each cut and make each cut only as far as the peg enters the seat.

Now put the legs into a bar clamp as in step 7 and clamp that clamp into a vise as shown. Shape the legs with a spokeshave or a drawknife as shown. You can smooth them down nicely if you want, but I left mine with a rough, hand-worked look. That's a large part of the charm of this type of project.

Make tapered wedges to fit the slots in the legs by tapering thin stock with a block plane or with a belt sander. Size the wedges so that they must be driven in with a hammer to expand the leg ends enough to tighten the joint. Put glue all over every surface that contacts other wood, assemble, and drive in the wedges. Clean off any excess glue with hot water and a rag while the glue is wet.

Once the glue is dry, cut off the tops of the legs and wedges as in step 8. Pare down the ends flush with the seat with a sharp chisel. Sand to your satisfaction, then rub down with a wipe-on oil finish for a natural look (see step 9 for finished stool).

Knickknack Shelf

Basic Game Plan

Tools: A table saw, a saber saw or a band saw, and a router

Materials: 1x lumber and ¼-inch hardwood veneer plywood

Time: One or two days

Parts Cut List: Knickknack Shelf (see figure 33)

Sides—2, ¾ x 4 x 24

Shelves—4, ¾ x 3¾ x 20

Top—1, ¾ x 1½ x 20

Plywood back—1, ¼ x 20 x 24

Here's a versatile design that you can use for a variety of purposes. As a simple knickknack shelf, this will look nice on the wall to hold small, pretty objects. Place thin strips across the front of each shelf, and you have a spice rack for the kitchen. Cut some simple dadoes in the shelves to fit sliding glass doors, and you have a bathroom medicine chest. This is a quick weekend project for which you'll only need a table saw, saber saw (or band saw), and a router.

I used oak because I think it's attractive, but any wood would be fine. Begin the project by ripping the parts to width and then cutting them to length. You can avoid having to do any thickness planing by using surfaced 1x stock, which is really close to ¾ inch in thickness. Be careful to choose straight stock or pick your pieces from straight sections of stock that, overall, might be crooked. If you choose to do the sliding glass doors and the shelves are not straight, the sliding

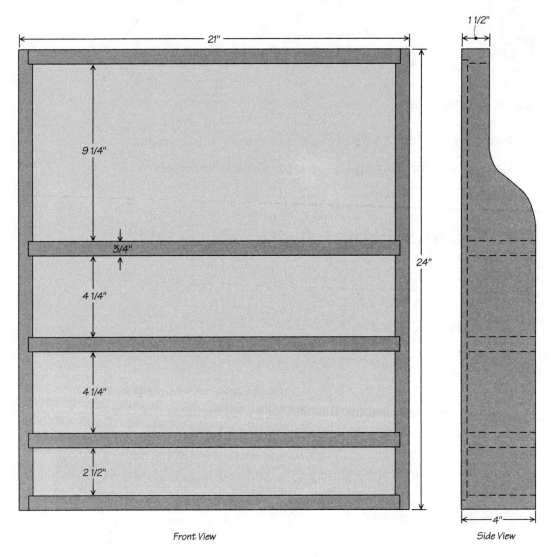

Front View   Side View

Figure 33.   Cut list for Knickknack Shelf.

glass doors could bind as they move, but a small amount of curve in the shelves (less than 1/16 inch) won't cause a problem.

Cut a rabbet on the inside back of the vertical sides into which the plywood back will fit, using a dado set on the table saw as in step 1. The parts list calls for 1/4-inch plywood, so of course the rabbet must be 1/4 inch deep so that the plywood flushes out to the back of the sides. However, you might have another thickness of plywood on hand that you would prefer to use up rather than buying a new sheet of 1/4-inch ply; if so, cut the rabbet to the appropriate depth to fit the plywood you have. Note that if you do use a different thickness of plywood, the widths of all the shelves and the top piece must be changed accordingly. The width of these pieces begins at the front of the plywood, so if the plywood and its rabbet are greater, the width of the shelves and the top piece must be less. The alternative, if you want the shelves to remain at the given width, is to make the sides wider to accommodate the thicker plywood.

Set up the dado set on the table saw to cut a dado equal in width to the thickness of the shelves. Getting a match between cut and thickness like this can sometimes be a challenge because often the stock is not at an exact measure; mine was at just under 13/16 inch, and my dado cutter cuts a maximum thickness of 3/4 inch. So, to widen the cut just a bit, I cut out paper shims in the shape of washers to go between the cutters in my set of stacking dado cutters. I had to do only a few tests with different numbers of shims before I got a good fit.

Put the miter fence onto the saw and set it at 90 degrees. Now set the rip fence so that it is just next to, but not touching, the dado set as in step 2. Use this setup to cut the rabbets in the top and the bottom of the sides for the bottom shelf and top piece. Set the depth of cut to 1/4 inch.

Step 1. Use a dado cutter to make a rabbet in the back of the sides for the plywood back to fit into. You could just as easily make this rabbet with two cuts with a combination blade.

Step 2. For the dadoes in the sides, you can use a dado cutter as shown or make two cuts for each dado with a combination blade and then knock out the waste in the middle with a chisel.

Step 3. Duplicate the dado setup to make the middle shelf dadoes by simply moving the table saw fence farther away.

Step 4. Sidebar: Cut the glass door dadoes in the leading edges of the shelves above and below the doors. Note that the dadoes above the doors must be deeper so that the doors can be cantilevered in.

Move the rip fence away from the dado set 2½ inches plus the actual width of cut of the dado itself so that when you cut the dadoes for the second shelf the distance between dadoes will be

Making Grooves for Glass Doors

By making a few careful dado cuts in the shelves, you can place pieces of glass within for sliding doors. The glass doors slide along shallow dadoes in the upper faces of the second and third shelves (counting from the bottom) and in dadoes in the lower faces of the third and fourth shelves. The dadoes for the top of the glass doors must be much deeper than the dadoes for the bottom of the doors so that the glass can be installed after the cabinet is made. Install the glass by placing it in the top dado as far as it will go so that the bottom of the glass fits over the outer lip of the lower dado and then drops into this dado as in step 4. You could lock the glass into the cabinet by making dadoes just large enough to fit the glass and install the glass when you assemble the cabinet; however, in this case, if a piece of glass breaks, you can't replace it without disassembling the whole cabinet.

Get your glass from a local stained-glass and leaded-glass maker and ask them to cut the four pieces to 4⁷⁄₁₆ by 10 inches. Cutting glass is not as precise as cutting wood, so it is possible that the actual height of the glass will vary by as much as ¹⁄₁₆ inch or more. For this reason, it's best to have the glass in hand before you make the dado cuts for the glass to fit into so that you can cut the dadoes to fit what you actually have. Having the glass in hand also shows you the thickness of the glass. Irregular glass with attractive wavy patterns comes in varying thicknesses. Your dadoes for the glass must be at least ¹⁄₁₆ inch wider than the thickness of the glass because when the glass is installed, it's placed in the upper dado at an angle and thus won't fit if the dado width equals the glass thickness.

In any case, your dado cutter probably won't cut a dado less than ¼ inch wide anyway, so if your glass is ³⁄₁₆ inch or less, use a ¼-inch dado. Cut the dadoes for the glass in the second, third, and fourth shelves as shown in steps 4 and 5.

2½ inches. Be sure to carefully orient each of the sides before making the cut so that you indeed make each cut on the bottom of each side. Both sides look exactly the same up to this point, having rabbets only on the top and bottom and a rabbet on the inside rear.

Again move the rip fence away from the dado, this time 4¼ inches plus the width of the dado cut as in step 3. Cut the dadoes for the third shelf, then again move the rip fence 4¼ inches plus the dado width to cut the dadoes for the fourth shelf.

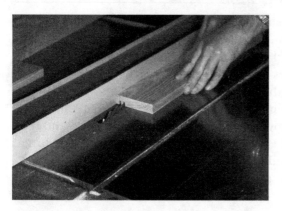

Step 5. Sidebar: Duplicate the dadoes for the second door. Just move the table saw fence over so that there is about ¼ inch between the dadoes and make the cuts again.

Step 6. After you cut out the shape of the curves on the sides, smooth the rough surface with a sanding drum on the drill press or mount the sanding drum in a hand drill. A third alternative is to wrap sandpaper around a dowel and hand sand the inside curve.

Step 7. A bearing-guided bit in a router table is a handy way to put a round-over on the outside edges of the sides, top, and bottom. You might choose another shape for the edge, but be careful that it's not so deep that the bearing dips into the dadoes cut for the shelves.

Step 8. Use nails to fasten the pieces together. Use a finish nail set to drive the nails below the wood surface, then putty over the holes.

Trace the curve shape shown in the drawing onto one of the two sides. The exact radii and their location doesn't really matter; for tracing curves like this, I usually just look for any available round object, such as a small paint can. Cut out this shape with a saber saw or band saw and then use this piece to trace the same shape onto the other side. This way they will be duplicates. Smooth the shapes of these curves with a sanding drum (available at most hardware stores), which makes this task go faster. Step 6 shows one mounted in a drill press, but you could mount it in a hand drill and clamp the piece in a vise for sanding. Either way, first use a heavy grit to smooth the saber saw cut and even out the lines, then use a finer grit to clean it up.

Mount a round-over bit in your router or router table as shown in step 7 and round over the outside edges of the two sides. Do the front edge and the top and the bottom, but not the back.

Glue and nail the shelves to the sides as in step 8. If you are using a hardwood, it is best to predrill for the nails with a small bit to prevent the wood from splitting. Putty over the nail holes with an appropriately colored filler.

Finally, cut the plywood back to size to fit in the dadoes on the sides and glue and nail it in as in step 9. If the cabinet will be used as a medicine chest close to a shower and will be subjected to a lot of moisture, use a finish that is waterproof to provide a good barrier to moisture absorption by the wood. Shellac is not moisture resistant, nor is an oil finish the best moisture barrier. Polyurethane varnish is a good choice in this case. Give it several coats and sand lightly with 320 grit between coats to smooth the finish (see step 10 for finished cabinet).

This cabinet is light enough that it isn't necessary to secure it to the studs in the wall. As long as you don't use it to store your prized collection of 50-pound fishing weights, you can just attach it to the drywall or plaster with mollies, which your local hardware store clerk will be happy to familiarize you with.

Step 9. Trim the plywood back to the right size to fit in the rabbets along the perimeter. Small finish nails tack it down.

Step 10. The finished Knickknack Shelf.

 ## Revolving Bookcase

Basic Game Plan

Tools: A table saw (a router and stationary sander will speed things up)

Materials: 1x lumber and "lazy Susan" hardware

Time: Four or five days

Parts Cut List: Revolving Bookcase (all ¾-inch stock) (see figure 34)

Bars—10, 1¼ x 44¼

Frame—8, 2 x 18

Frame—8, 2 x 14

Panels—4, 13½ x 14½

Dividers—3, 2 x 17½

Crosspieces—2, 3 x 18

Feet—4, 3 x 3

Base fascia and supports—6, 1½ x 16 (approx.)

Here's a space-saver bookcase that can be located in a corner of a room or in the middle of a room next to other furniture. It holds as many books as a wall unit that's about 4 feet square but requires little if any wall space. The revolving feature is easy for you to make by using "lazy Susan" swivel hardware, which is available at most hardware stores. This is a table saw project and can be done with this tool alone, although a router table and a stationary sander can help. But all the joinery is done with a dado blade or your regular combination blade at the table saw.

Use 1x lumber throughout. I used cherry, but suit yourself. Be careful to pick straight stock for the shelf frames and their panels. The more twisted and bowed these pieces are, the harder it is to edge glue the panels and to cut accurate, tight-fitting grooves and tenons in the frames. However, the 10 vertical posts don't need to be as straight because they get only simple dadoes that are cut across them to fit the panels. When you screw the vertical posts to the shelves, you can pull them straight if they're a bit off.

Edge glue the stock for the panels as in step 1. Before you edge glue the parts, you need to ensure that the edges to be glued are straight enough to mate well for a good glue bond. If you don't have a jointer for this, run the pieces through the table saw numerous times, flipping them between cuts so that you alternately rip one edge and then the other. Each time take off only 1⁄16 inch. After doing this three or four times, variations will cancel out, and the edges will be straighter. Note that the piece shown is long enough for two panels.

While these dry, cut out your parts for the shelf frames. To make the frame parts uniform in length, it's a good idea to use a

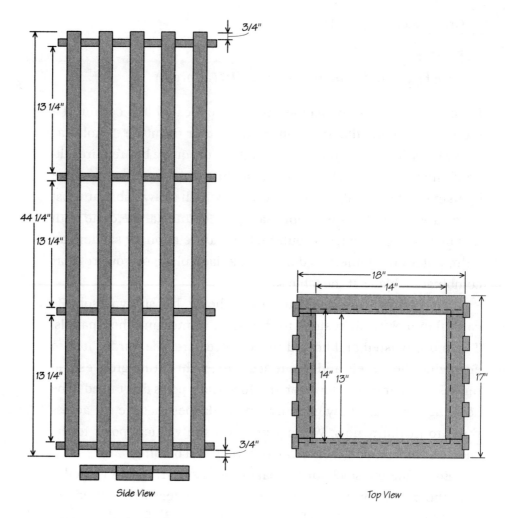

Figure 34.   Cut list for Revolving Bookcase.

stop block when cutting the pieces to length at the radial arm saw
as in step 2. The same idea can be applied to a cutoff box on the
table saw.

Cut a groove along the inside edge of all the frame parts. Make
the groove ¼ inch wide and ½ inch deep. This groove will double as
an open mortise for the tenons on the frame rails, so for a strong

joint at the tenon you need at least ½ inch depth on the groove. Make the groove with a ¼-inch dado setup in the table saw or use your combination blade as in step 3. Set the blade ¼ inch from the fence and run each part twice, alternately on both faces of the part. This method results in a well-centered groove.

Make the tenons on the rail ends with a dado blade and the miter fence as in step 4. Clamp a stop block to the fence before the blade as shown to locate the end of the rail prior to the cut. This is necessary so that you can locate the fence away from the dado blade. Screw a backup fence to the miter gauge as shown to back up the cut as the dado blade comes through, preventing tear-out on the rails. Butt the rail end against the stop block, hold the rail firmly against the miter fence, and push the blade through.

Note that the height of the dado blade determines the thickness of the tenon. Make your first cuts on a test piece and adjust until the tenons have a snug fit in the grooves.

If you used yellow aliphatic woodworker's glue when you edge glued the panels, they can come out of clamps after about an hour, perhaps just as you finish the tenons. Size the panels, then use a dado in the table saw to cut tongues along their edges to fit the grooves inside the frames. Set the fence at ½ inch from the dado and run the parts on edge as in step 5. Adjust the distance of the fence from the blade so that you get a tongue that fits the groove loosely but without rattling around. Do not run your fingers directly over the blade within 6 inches of it; rather, move your hand around the blade to the front of the panel halfway through the cut.

Make this cut on both sides of the panel along each edge. First make the cuts on the end grain and then on the edges along the grain to clean up tear-out on the rear end of the end grain cuts. On the end grain cuts, set the blade height to ¼ inch, but for the cuts along the grain, make it almost $\frac{1}{16}$ inch higher. This will give the panel room to expand in the frame as moisture increases so that the panel doesn't push the frame apart (as it could do).

Glue up the frames as in step 6. Don't put glue onto the panel tongues—only on the tenons and in the grooves where the tenons fit. You must allow the panels to move, which they can't do if

Step 1. Use bar clamps to edge glue as many boards as you need to so that you get the width required for the shelf panels.

Step 2. It's easier to make pieces the same length when you use a stop block to register the location of the piece to the blade. You could simply clamp a piece of wood to the fence rather than use the commercial variety shown here.

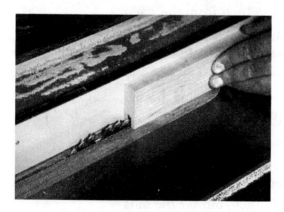

Step 3. Cut grooves in the edges of the frame parts at the table saw as shown with a dado blade or combination blade.

Step 4. This tenoning setup with a miter gauge and dado cutter makes tenoning fast. Note that the distance of the fence from the blade determines the tenon length.

they're glued in. First pull the stiles onto the rails with bar clamps, then put C-clamps onto the joints as shown to squeeze the groove walls onto the tenons. Locate the outer rail edges flush with the ends of the stiles. After the C-clamps are in place, remove the bar clamps and use them for the next frame.

While the frames dry, get out the horizontal bars. Don't be afraid to expand or lessen the width of these parts as you desire or as your stock dictates. They must, however, be uniform in length so that all the frames will line up well in the dadoes when the piece is assembled.

Cut the dadoes in the horizontal bars with a dado cutter in the table saw as in step 7. The width of the dado should match the thickness of the frame parts for a snug fit. You can make small adjustments in the exact width of a stacking dado cutter by making donut-shaped shims out of paper and placing these between the layers of the stack. Raise the dado cutter to ¼ inch above the table. You can make these cuts with your combination blade at the saw by sliding the part back and forth to get the width of each dado cut. If you do this, screw a long backup fence to the miter gauge so that you can put a stop block on the other end of the bars to limit the width of the dado cut.

Screw a backup fence to the miter gauge as you did for the tenons to back up the cut and prevent tear-out. Use the rip fence as an end stop to locate the dadoes along the length of the bars as in step 8. Note that the end dadoes are equidistant from the bar ends, so you can use the same setup to cut both end dadoes on each bar simply by flipping the bar. The two middle dadoes on each bar are also equidistant from the ends, so the same principle applies.

Round over the ends of the horizontal bars with coarse sandpaper on a sanding block or with a stationary sander as in step 9. Round over the long edges of the bars with a ⅛-inch bearing-guided round-over bit in the router mounted in a router table or by hand sanding with a sanding block.

Pull the frames out of the clamps and level out the ends of the stiles with the rails using a block plane (or sanding block) as in step 10. First get rid of any residual glue clumps with a scraper or bang

Step 5. Cut a groove along the edge of the panels with this setup. Keep your fingers away from the blade. You can build a higher fence for the saw to make this operation easier.

Step 6. Before you glue the frames, try dry fitting them with no glue to be sure that everything lines up. Then take it apart and glue it for real.

Step 7. Cut dadoes near the ends of the vertical bars with this dado setup and the miter gauge.

Step 8. Move the table saw fence a bit farther away, and you're all set to cut the dadoes in the middle.

them off with a dull chisel. The glue in the joint will still be enough to dull your block plane iron a bit, but this is a quick and easy way to level the joint, and you can hone the iron again later.

Cut dadoes for the crisscross base using a setup similar to that for the dadoes on the horizontal bars. Make the depth of these cuts halfway through the thickness of the parts so that when they join, the outer faces will be flush. Cut 3-inch-square footpads for the cross and glue all the parts together as in step 11. When you run out of clamps, round over the exposed edges as you did before.

Assemble the shelves and bars using screws as in step 12. Place the bars on the sides of the shelf frames that have the exposed joints and cover the joint with the outer bar. Locate the screw hole for these bars closer to the middle of the entire shelf so that the hole doesn't land on the joint.

When you have a lot of screws to apply, use a tapered drill bit with a countersink. The tapered bit makes a hole that matches the screw shank well, and the countersink makes a good seat for a wooden plug to hide the screw head. Use a variable-speed drill with a Phillips driver tip to seat the screws (52 screws by hand is hard on the wrists!).

Use a bar clamp to hold the bars in place while they are screwed down. Once all the bars are in place, secure the central dividers in a similar fashion as in step 13. Cut them to fit snugly between the middle bars and clamp them in place as shown, locating them midway between the upper and the lower shelves.

Cut out plugs for all the holes using a plug cutter at the drill press or with a hand drill at the vise. Glue the plugs into the holes and clean away excess glue while the glue is wet. When the glue dries, level the plugs with a sharp chisel or block plane.

Cut out four perimeter pieces for the base and screw them to the bottom of the lower shelf frame as in step 14. Miter the corners

Step 9. A stationary strap sander is really handy for rounding over the ends of things.

Step 10. Use a block plane or sandpaper and a sanding block to level out the joints on the shelves after gluing them up.

Step 11. Now you see what they mean when they say you can never have enough clamps. By applying a little pressure with a lot of clamps, you evenly distribute the pressure over each joint to ensure a good bond over a broad area.

Step 12. Screw the bars to the shelves as shown. Be sure that the screw holes are large enough that the screws don't shear off as you sink them but not so large that the screws don't grab. Doing tests in scrap wood is a good idea.

on these pieces and cut them long enough that, when in place, ½ inch of the inside of the frame is still exposed.

Lazy Susan swivel hardware comes in a variety of sizes, but don't use one smaller than 12 inches in diameter because the book-

Step 13. Locate the divider bars as shown. These bars will keep your books from being pushed in too far.

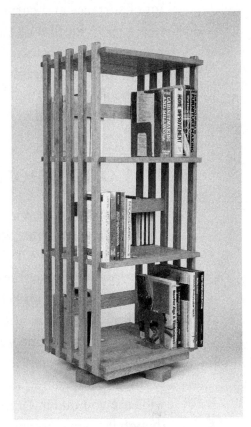

Step 14. Note the two pieces attached to the bottom shelf that the lazy Susan hardware is screwed to. They must be located to line up with the screw holes in the hardware plate.

Step 15. The finished Bookcase.

case will rock back and forth. Cut two support pieces for the lazy Susan and screw these to the bottom of the lower panel. Cut them long enough to extend over the frame of the lower shelf as well so that weight is transferred to the support pieces by the frame as well as by the panel itself.

Screw the lazy Susan to the crosspieces first, then screw that assembly to the support pieces already in place on the cabinet.

A wipe-on oil finish is a good choice for a piece like this that has a lot of surfaces and corners that would be hard to hit with a brush (see step 15 for finished bookcase). Also, you don't need the kind of protection that a brush-on polyurethane provides because the bookcase is unlikely to get wet. However, you'll be better off filling it up with books before the kids get to it, or they might think that you made this new "spin toy" just for them.

In Conclusion

Once you've tried a few of these projects, you'll begin to feel more confident with your woodworking skills. These projects are grounded in basic skills: using the table saw, rounding over with a router, and cutting curves with a band saw. Using these skills, you can proceed with more complicated projects of your choosing. Perhaps you have a specific project in mind that you want to do but that still seems beyond your skill level. Look in woodworking magazines or project books for similar but easier projects that use some of the techniques required for that project. Making a few of these will ease you into it. Or just jump in and try that tough project even though it seems too hard! You might make some bad mistakes, but you'll learn so much from those mistakes that your second and third tries will be far better. Remember, to make an omelet, you have to break some eggs!

Your Crafts Vision

NOW THAT YOU'VE HAD SOME FUN and have learned a bit about woodworking, you'll want to take a serious look at the direction you might take from there. You should consider both the aesthetic and the pragmatic aspects of your decision. You must couple your vision of the design style that you really want to pursue with the practical economics that will be required for you to continue in your craft.

What is your aesthetic vision? Most artists take inspiration from the past, which they use to help formulate their own vision. Perhaps you like the flowing lines of art nouveau furniture from the early 1900s or the elegant simplicity of Shaker designs. Maybe the extravagant excess of baroque ornamentation tickles your fancy, or perhaps the craftsman ethic of basic, quality designs is your point of departure. You might decide to pursue the goal of turning a perfectly shaped bowl or sculpture on the lathe or using a chainsaw to carve out life-size statues from tree trunks. Any large library has a section on furniture and other woodworking design that includes books full of pictures from different periods and of different styles.

How do you approach selling your projects? Perhaps you have wanted to carry your craft beyond just making things for yourself and your friends and begin to sell them to bring in extra earnings in your spare time. Woodworking lends itself well to this sort of part-time business endeavor for several reasons. Unlike a service business, in which people will be counting on you to be available to help them at the time they most need it, you can do your woodworking entirely on your schedule. If you want or need to, you can drop it for weeks or months and get back at it when it suits you. That's not the case if you have gallery owners calling and demanding more pieces to sell. In addition, a one-person woodworking operation can require only a small amount of space and resources—just a corner of a garage and at least a few simple tools. Thus, you might not need to pay a lot of rent or borrow money to begin producing items for sale.

After I had been woodworking in a commercial shop for a couple of years, I became curious about how I might apply my skills to make a little money on the side, but I had no idea where to start. I saw the many avenues that other woodworkers had taken to sell their work. Local craft shows always featured woodworkers with a wide variety of wooden items. High-end galleries were selling wonderfully made pieces of furniture. Some woodworkers set up high-output production facilities to make a large volume of lower-priced items for sale to a variety of retail and wholesale customers. I had friends whose work seemed to come simply from word of mouth as they went from one type of job to something completely different, depending on the needs of their clients.

I made my first foray into doing woodworking as a business by making ten jewelry boxes in a small production run using highly figured local woods. I proudly took several of them to a local fine crafts gallery to show the owners, only to be disappointed when they told me that they would not take them. It seems that I hadn't wiped all the glue out of the corners on the inside—a beginner's mistake. Undaunted, I sold several of those boxes to friends and gave others away. Next I filled a specialized niche—small wall-mounted display

Woodworking Niches

The following list outlines niches that other woodworkers have occupied and that you might consider for yourself.

- Local craft shows, knickknack galleries, and craft malls: homespun items, such as lawn ornaments, whirligigs, children's toys, kitchen utensils, and simple furniture (e.g., three-peg stools). At these venues, you can sell inexpensive items that look nice, such as simple boxes, small carvings, lathe-turned candlesticks, simple rustic furniture, or household items, such as handheld mirrors, serving trays, or knickknack shelves.

- National craft shows and high-end galleries: fine furniture with your own design, very elaborate boxes, or sculptural objects, such as turnings or carvings. The emphasis here is on art and your creative vision.

- Word of mouth: small-scale custom built-in cabinets for friends, neighbors, and others, or custom-designed furniture to meet their specific needs.

cases for showing glass paperweights that a local merchant needed. However, because this customer's needs were so limited, I decided to continue writing articles for woodworking magazines and selling to a local gallery the pieces I had made.

Perhaps the most important thing to consider as you take on woodworking as a business, in any form, is what niche you can or want to fill. You can go in so many different directions, but you need to find one that is right for you. Perhaps all you want is to sell enough projects to cover the costs that you have incurred starting your craft. Perhaps you want to develop your craft into a second income. You also need to consider the type of woodworking you can do best. You might have certain resources, tools, or skills available to you that give you a certain advantage in a given area of woodworking, such as a beautiful wood that grows locally or your own skills with a lathe or other tool. Maybe you're just so determined to do one thing very well that you carve out your niche in that direction solely to realize your dream.

Identifying Your Dreams and Goals

Use the following list of questions to help you identify the direction that is best for you. But don't limit yourself simply to what might seem to make the most "economic sense." Maybe it's more important to you to fulfill a certain dream with an uncertain financial future than to follow a path that seems to have more tangible rewards. You'll do better at something that you want to do than you'll do at something that was your second choice.

- Without any limitations or constraints of "common sense," imagine for a moment what your ideal woodworking career would look like. Where do you really want to be with your craft?
- Realistically, what would it take, from where you are now, to get to where you want to be?
- Are you willing and able to take the time it might take to achieve your goals?
- Are you willing and able to make the sacrifices required to achieve your goals? If not, is there some way you can realize some of your goals within the time allotment you have given yourself while considering the sacrifices required?

Now let's take a look at your craft goals from a financial perspective.

- Is what you want to do expensive in terms of machinery and materials? Can you afford to get started in this right away? If not, can you gradually obtain the required machinery and materials over time, doing something simpler in the meantime?

- Do you want to make just enough money to pay for your craft, or do you want your craft to provide you with a second income? If you want your craft to provide a second income, are you willing to make the hard business decisions and sacrifices that might be required to make your business work?

Finally, let's take a look at how your particular circumstances might help you find a unique niche.

- Are unique woods or other materials available locally that might help you sell your woodworking?
- Do you have a unique marketing outlet that you can take advantage of to help you sell your work? Do you have a particular skill or specialized tool that will allow you to provide a desirable product?

Your niche, then, is not just a situation that is most economically viable or practical. It's a situation that you enjoy, one that means something to you, and which is economically viable and practical. Be adaptable, keep looking around you for alternatives and you'll find the niche that suits you.

Part Two

For Profit

Profiting from Your Talent

ENTERING INTO YOUR NEW business venture might seem just as challenging as when you entered your woodworking craft for the first time. There are all levels of being in business, just as there are all levels of being a craftsperson. By "being in business," I mean anything from selling a few things to your neighbors to becoming a serious production woodworker selling a steady volume of work. However you do it, the extent to which you're serious about selling your woodworking should equal the extent to which you investigate the necessities of being in business. A well-organized business will be more profitable, if only because you spend less time spinning your wheels and more time producing things to sell. In addition, by planning ahead you can save money on materials and avoid other costs that cut into your profits.

This section of the book gives you information on the various aspects of operating a woodworking business. We'll look at how to set a price on your woodworking and where to sell it. Then we'll look at advertising and publicity, and finally we'll look at basic business practices, laws, and regulations, as well as copyright issues. As you consider each of these subjects, you'll need to think about the extent to which you need to get involved with each, given how far you want to go with your craft business. If you simply plan to make a few things here and there and not go into it in a big way, you'll still find things here to help you. If you do plan to go into your business in a big way, carefully investigating all aspects of your business venture can make the difference between the longevity or failure of your business.

Overcoming Hesitations

You might have some hesitations about entering into business with your craft. When I began selling things, I wondered whether I was good enough to be selling woodworking along with all the "big boys" making beautiful pieces of furniture. What I found was that my self-confidence expanded as my experience did. I made a few mistakes here and there (such as not wiping off the glue inside a jewelry box), but you learn from your mistakes. If you don't want to jump into the deep end before you know how to swim, jump into the shallow end and just get your feet wet. Start small and work up from there. Bite off a workable chunk before you think about the bigger task that now seems unapproachable.

> **Handy Hint**
>
> Filing a Schedule C tax form for your business, small though it might be, can save you money on your taxes by allowing you to deduct expenses.

This means starting small with your projects until you are thoroughly familiar with everything that's entailed in making and selling items. If I had it to do over again, I would have made only two or three of those jewelry boxes my first time instead of ten. Then I would have found out what the gallery owners wanted before I had spent so much time making them. In addition, I would have known how long it takes to make them, and I could have simplified the design or made it more complicated, depending on what I thought customers might like and might be willing to pay.

You might have some hesitation about entering into business for yourself. After all, you might have become involved with the craft for creative purposes, not to be an accountant. Some woodworkers who expand their business by adding employees end up being managers who are divorced from the creative part of their craft. That's another reason to start small and work up from there. You don't need to take on all the business aspects of your craft-as-business all at once, especially if you have a small operation at the start. Spend most of your time practicing your craft. After all, the business part is there because of the craft, not vice versa.

Self-Confidence in Your Work

Projecting self-confidence in yourself and your work is an important part of selling your woodworking. In some cases, the craftspeople are selling themselves as much as they are selling the work itself because people want to know that they're getting something from a truly experienced woodworker. People respect someone who can make an object well by themselves with only their skill, knowledge, and experience, and they want to know that you are that type of person. Tell people what you have done and how you care about your craft, and be sure to show them photos of the work you have completed. These things will help instill confidence in others about your abilities.

> **Handy Hint**
>
> Photography might seem like too great an expense for a very small business, but having good pictures can make the difference between selling pieces and not selling them.

Because people look to independent woodworkers such as you for unique design ideas that they don't see in factory-made furniture, having your own identifiable style is a real asset to selling your woodworking. Although some woodworkers make whatever style the customer orders, people often look to the woodworker to supply a design idea that they hadn't seen before. When people approach your booth at a craft fair or you approach a gallery owner, they will assess your individual style and decide whether it's for them. Some will like it, and others won't, but those who do will seek you out again because they know what they can get from you or will recommend you to others as having a recognizable style. You need the self-confidence to make your own identifiable style and to tell people about it.

The Broad Product Base of Woodworking

Since prehistoric times, woodworking has supplied people with the necessary implements of daily living. This was done out of convenience and necessity. In the past, wood was a readily available material that could easily be worked into the implements people needed. But with affluence came luxury, and many of these items were embellished with beautiful designs and fine finishes to satisfy the desires of the users. This dual nature—utility and beauty—is still present in all woodworking today. Although many wood items still have a strictly utilitarian purpose (e.g., tool handles and simple kitchen implements, such as wooden spoons), most woodworking must incorporate in its design some degree of aesthetics to be at all viable on the market. People want the implements they use on a daily basis to look good. Because so many of the implements that are foisted on us are made of cheap plastic, wood has come to symbolize quality, even though it's just as possible to make something cheap out of wood as anything else. Indeed, wood was the plastic of the past, the most easily manipulated material. But now people demand that it be presented to them in a way that shows off the wood's beauty and that speaks about the quality of hand craftsmanship.

Wherever the warm appearance of wood can be used to good advantage in the products we use, people like to see it. I recently saw a computer keyboard and mouse made of wood—far more expensive than the plastic variety but so much better looking. Making a computer mouse out of wood is kind of like stuffing a square peg in a round hole because it's far easier to make the push buttons from plastic. But here's where aesthetics overrule utility and we find that some people are willing to pay the far higher price for the more beautiful item.

This issue of aesthetics overruling utility is key to the small independent woodworker who's trying to decide which items to produce. The large factories have you beat hands down when it comes to producing wood items that are mainly functional, such as simple wood chairs, mass-produced cutting boards, and inexpensive bureaus. But as soon as you take any of those items and make them more beautiful and better crafted than those that the factories produce, your work will appeal to the customer for whom aesthetics overrule utility. As you look at the broad field of possible items to make, always think about how you can make that item different, more beautiful, and more unique and, at the same time, useful for a given purpose.

Here are a few examples. One woodworker thought that he could produce children's building blocks efficiently on a production basis. However, he could not compete with overseas manufacturers even though he worked quickly. Thus, he built his blocks using hardwoods of many different types instead of one. These blocks sold by word of mouth (even though they cost more) because customers liked their appearance.

A turner in Texas specializes in turning mesquite, a tree whose lumber is not widely marketed. But this wood is very beautiful, being distinct from other common hardwoods, so his bowls and other objects are unique as well.

I have a friend who used to sell small wooden clocks in craft shows. He would glue colored pieces of countertop laminates to the faces of the clocks to give them the appearance of marble and other materials.

Another friend of mine lives in a woodland setting that includes many eucalyptus trees. Although eucalyptus is notoriously hard to work with because it splits and twists when it dries, my friend uses it to carve walking sticks that include sculpted figures of lizards, frogs, and other animals climbing up the stick. The splits become part of the character of the sticks and don't compromise the structural integrity of the piece as they do in furniture. He has sold many of these in local galleries and has had photos of them published in a national woodworking magazine.

Venues for Selling Your Woodworking

Turning your projects into cash necessarily entails finding the right way to reach the right customers. A beautiful and expensive piece of furniture placed in a shop that caters to tourists buying knickknacks will probably not sell because it isn't being exposed to the right group of people. You need to assess the different venues available to you to find those that are sought out by the customers you're seeking. Thus you establish a mutually beneficial relationship. If the local venues are not appropriate, you might need to look farther away, and although this can be costly, it might be the only way for you to sell your work. Let's take a brief look at the major types of venues.

- Galleries and shops. Some shops depend solely on passers-by who buy small items on a whim; other high-end galleries cater to customers who think long and hard before buying expensive, well-made items.
- Craft malls. These are large buildings that house many small shops and galleries or that consist of one large gallery with many small areas devoted to different crafts.

You'll find antiques, quilts, fiber arts, ceramics, imported items, and any kind of knickknack you can think of. Simpler, less expensive woodworking items generally will sell here.

- Craft shows. Local craft shows and craft sections at county fairs cater to people looking for inexpensive local "color." National craft shows focus on quality art and crafts that are on the cutting edge of contemporary crafts.

- Internet. The Internet lets you put a picture of your work where anyone with a computer and a modem can see it. You can set up your own Web site or place your photos on a gallery's or craft mall's site. Many such sites exist, and more are always being created.

- Wholesalers. You only want to consider wholesaling if you are set up for production woodworking in a big way because you must produce a very high volume of items. Some craft shows devote certain days only to wholesale customers who contract with makers to supply quantities of products to them for sale in whatever outlets the wholesaler uses. Larger factories have their own wholesaling networks, but you can't do that without ten employees on an assembly line.

Production Woodworking Techniques

The business of woodworking has changed a great deal over the last few hundred years. Gone are the days when the king of France dictated the furniture style of the day and employed numerous woodworkers and designers all endeavoring to place the king's indelible

Did you know???

Internet sales are projected to exceed $220 billion by the year 2001.

stamp on the history of woodworking design. Changing politics, social patterns, and the industrial revolution have given us a large woodworking industry that is based mainly on large production facilities that make high quantities of products in the same design. Their production techniques allow these factories to bring the price of their work down but don't allow them to do quirky, unique kinds of woodworking or to custom tailor what they do to the specific needs of customers.

That's where you come in, as you can make something different that the factories just can't handle. At the same time, though, many independent woodworkers such as yourself find it necessary to use production techniques to a certain extent in order to make a reasonable amount of money for their woodworking. Any way you look at it, most woodworking is labor intensive. In the final analysis, to make a reasonable hourly rate, you must do all you can to reduce the hours it takes to produce each piece of work. Some woodworkers feel that they lose their sense of creativity as they become more production oriented, and none of us wants to feel as though we're working on an assembly line. But some of the best examples of work from independent woodworkers are those that combine both elements: creativity and production. Find a unique design that incorporates your individual flourish but that can be made in whole or in part using efficient methods that reduce the time it takes to make each piece.

For example, one woodworker likes to build large pieces of furniture out of thick chunks of wood that he carves deeply with long lines in deep relief. Doing all this by hand took too long, so he built an elaborate router-duplicating machine that allowed him to carve wood with a router by following a duplicate template. He makes the templates out of stiff foam, which is easy to carve. Once he has the template, he can carve as many parts as he wants much faster than doing it by hand. You probably won't build an elaborate router duplicator like he did, but you can build or buy small, simple jigs (like dovetail jigs) that will increase your production speed.

Sometimes when I wander through a gallery or shop, I'll find small boxes produced by a local woodworker. I'm always pleased to see the different designs and beautiful woods used in these. On closer inspection, I always find that they're made with simple, fast

Jewelry Box

Three-Peg Stool

Knickknack Shelf

Revolving Bookcase

joinery techniques, such as a tongue-and-groove joint that can be done on a table saw or even a simple rabbet joint nailed together. Such simple joinery in boxes is structurally adequate for what boxes must do and does not visually detract from the piece, and the work can be done quickly enough to make the price you want to charge reasonable.

A friend of mine was commissioned by a customer to build a very fancy veneered cigar box. To do this, he had to get a special veneer vacuum bag setup that was expensive and time consuming to learn to use. Instead of making only one box, he made six at once. Having learned the veneering techniques, he was able to do the additional veneering quickly. By word of mouth from his first customer, he sold the other boxes, which paid for the time and expense of the vacuum bag setup.

You can use this book to alter your manufacturing techniques if you choose. Finding the most efficient way to produce an item is an important part of being in business, so be open to changing your techniques for greater efficiency. Review the variety of techniques outlined in this book and think about how some of these could help you. You might find that by slightly altering your design, you can use different techniques that will make your operation much more efficient.

Let's take a general look at various production methods in the woodshop to see how some of these might be applied to your woodworking.

> ## Handy Hint
>
> Ultimately, you're competing against large production facilities whose price you can't beat, but you can offer better designs with more artistic flair that people will pay more for. But experience has shown that most people will not pay much more for this, so you must temper your artistic inclinations with a measure of practicality. Still, most people feel that this is a small price to pay for being an independent craftsperson.

What Is Production Manufacturing?

Production manufacturing means setting up a machine and/or jig to do a certain operation and then using this setup to produce a large number of parts for a large number of finished items. Let's use my little run of ten jewelry boxes as an example. When I set up the table saw to do the joinery on the box sides, I ran all the parts for the ten boxes through that setup. If I had set up the table saw ten separate times to do the joinery for ten separate boxes, I would have spent the time it takes to set up the machine nine more times. Thus, the average time needed per box to do the basic joinery was much less than the time needed to make one box. This is the time-saving advantage to using modern machinery, as in the past each piece had to be individually set up in a vise and worked by hand. With those methods, making ten pieces took about as much time per piece as making one.

But to do production manufacturing, even on the small scale of building ten jewelry boxes, your stock must be very uniform. Any pieces of wood that are of a different thickness than their neighbors or that are not straight will not go through the machine setup the same way as the others will. Thus, those pieces that are not uniform will have visible gaps, inferior joints, or other defects, and your efforts will be fruitless. For this reason, it's very important to be able to straighten your stock well because lumber, as you buy it, is not straight enough to go through machine setups well. To straighten stock, you need a jointer and a planer.

Straightening Stock

Begin straightening your stock by face jointing, which means putting the board down flat on the jointer table and jointing the broad face of it to make it flat. However, before you face joint each board, you should check to see how flat it is to begin with. If the board is much more than ⅛ inch out of flat over its length, you'll need to face joint off at least ⅛ inch to flatten it. This begins to seriously reduce the thickness of the piece, and you don't want to end up with pieces that are too thin. You can cut boards that are this bad or worse in half, with the result that the two boards will be much closer to flat than the long board they were at the start. Now you have less to remove on each board by face jointing to straighten the stock.

Of course, you can't cut pieces in half that need to be long. So always start by looking for your longest pieces. When you're in the lumberyard pulling stock, look for straight pieces from which you can get your long pieces. Get a few bowed ones too; you can cut those up for short parts. The folks at the yard will look at you sideways if you sort through and take only the straight stuff. Back at the shop, account for your long pieces and cut them close to length before you face joint. Now do all your face jointing on all the parts.

The next step is to plane the stock to thickness. After face jointing, the stock will vary in thickness because one face is flat and the other isn't. Your planer will bring all the stock to the same thickness, which a jointer can't do. But a planer can't straighten, and that's why you joint first. Place the flat face down on the planer table and let the planer shave off the top surface, leaving you with flat stock of uniform thickness.

Now you need to do the same thing with the edges of each piece that you did with the faces. Run each piece through the jointer on edge to get that edge nice and straight. Then rip all the stock to your desired widths on the table saw. Place the straight, jointed edge against the table saw fence and rip off that ugly rough edge. The result is called S4S (surfaced four sides) and is now ready to be cut to length and run through your machine setups.

Handy Hint

If you need to plane boards thinner than ¼ inch, use a backer board. Place the thin board on a thicker board and run the two through the planer at the same time. This way there is less chance that the thin board will break apart as it is cut.

Problems with Straightening Stock

All wood has some tension in it. When the wood dries and shrinks, some parts shrink more than others, resulting in a kind of isometric balance of forces. When you remove some of the wood by jointing, planing, or sawing, you remove one area of this force and so disrupt the balance of forces. The result is that some (but not all) of the boards move a bit after being cut. It's no fun to take a board with a freshly jointed straight edge to a table saw, rip it in half, and then find that the ripped piece is no longer straight. Well, as they say, you can't win for losing. But there are ways to minimize this problem.

The less wood you remove from each piece, the less chance you'll see bad movement. Pieces that are close to what they need to be from the start probably won't cause you a problem, but when you must rip wide boards, each into several skinny ones, you'll see movement on some of them. To deal with this, rough rip the wide board before you do anything else with it, ripping about ¼ inch over the final width. Then edge joint each piece and finish rip it to width. When you must greatly reduce the thickness of pieces at the jointer and planer, first face joint to remove half the waste. Then let the sticks sit overnight and check them the next day. Again face joint the ones that moved. Now plane all of them close to thickness and leave them for a while. Again inspect and face joint any that are laughing at you. Now finish plane to thickness.

Some Basic Production Setups

In the past, mortises for mortise-and-tenon joinery were cut by hand with a chisel and a hammer. This required experience and skill to end up with well-aligned mortise walls that accurately located the tenon. With machine setups, you use the accuracy of the tool or jig to create well-aligned mortises. A plunge router easily makes the plunge cut required to cut a mortise—you need only build a jig that holds the piece in proper alignment to the bit so that the mortise is properly located and its walls are parallel and perpendicular to the outside wood surfaces. A router-cut mortise will have round ends to it from the bit, which then leaves the problem of fitting a square tenon in a hole with round edges. Rather than trying to cut tenons with rounded edges, I just use a chisel to square up the mortises.

But that takes an extra step, and for machine production work you want to minimize the number of steps you take. Thus, a chisel mortiser in a drill press is a good alternative because it cuts square mortises to begin with.

Cutting tenons on the table saw is fairly fast and easy using a jig that holds the parts up on end so that you can push them by the blade. I like to set up on both the table saw and the band saw to cut tenons, with one tool cutting the faces and the other the shoulders. This way I don't have to break the setup on one machine to do the other operation.

> ### Handy Hint
>
> Get glue into dowel holes with a small disposable brush, or whittle a dowel down just under size so it fits loosely into the hole. Dip the dowel in glue and then place it in the hole.

Despite the use of such machine setups, mortise-and-tenon joinery is still very time-consuming to produce. Dowel joinery is much faster because you simply need to bore holes in each piece of wood. Boring holes in the edges of pieces is easy with a drill press, but what about boring holes in the ends of pieces? This is not so easy on a drill press, but it is possible. Boring machines were invented for this purpose, making hole boring in both edges and ends a snap. But for those of us who can't afford a boring machine, dowel boring on the ends of sticks using a dowel jig is the best alternative.

For production box joinery, you can set up on the table saw for finger joints, splined miter joints, and so on. Dovetail jigs with a router will make a very attractive joint rather quickly. It can take a while to set up one of these jigs, but once it's set up, you can cut the joinery for many boxes or drawers in just a day. People love dovetails because they symbolize quality wood craftsmanship.

You Can Do It!

Profiting from your woodworking is something that you can do, with a little effort, focus, and organization. You'll need to put in the hours to learn your craft and produce quality pieces. You'll need to focus on which venues are the best for you to sell your pieces, and then you must do what's necessary to get into those venues. And you'll need to organize your operation efficiently using production techniques to reduce your time. With a little spit and spunk, you'll make those initial sales—and repeat sales—that show that you're doing more than just having fun in your spare time!

Pricing Your Woodworking

IT'S VERY IMPORTANT IN ANY business to keep a close eye on time, costs, and the final return you receive for your work. This is especially true in woodworking, which is a very time-consuming endeavor. You simply can't continue to spend a great deal of time doing something that garners a limited return. By having an accurate means of reviewing the time and costs that you put into a project and then reflecting on how it sold and on the reactions you got to your work, you can make the best decisions for your future projects to make them more profitable. You might decide that your customers didn't really appreciate a certain time-consuming aspect of what you did, so that if you eliminate that aspect you can offer the piece at a lower price and still satisfy your customers.

The opposite might happen as well. You might find that people are not willing to buy your pieces at a low price because the pieces aren't fancy enough. In this case, you might sell more pieces more profitably at a higher price by adding time-consuming flourishes to the pieces, as this is what makes them sell. But you need to know how much time this will take and the expenses involved so that you can make a solid determination of how much time and resources you should devote to each project. You could simply guess without keeping records, but the degree of accuracy between guessing and keeping good records might be the difference between being profitable and not!

Controlling Costs

Wood and the other supplies that you need for your craft are expensive, and limiting their use is an important part of woodworking efficiency. Using your wood efficiently begins at the point of selecting your stock at the lumberyard. You want to choose boards that are of such a size that you can efficiently cut out what you need from them. If you need pieces that are all 4 feet long, choosing 6-foot-long sticks is inefficient because you'll cut 2 feet off every one of them.

How you get your pieces out of your stock is always a jigsaw puzzle, especially when you have many parts of different sizes. There are a million ways to do it, reshuffling the various combinations this way and that. You can stand there all day and ultimately find the most efficient combination, but how efficient are you really being if it takes you four hours to save two board feet? At some point you must simply go with your best inclination, making sure that you don't do anything grossly inefficient other than striving for the best use of your wood.

If you buy your hardware in bulk, you can save money per piece. But be sure to use all the pieces you buy because, if you lose or you don't use two or three out of a discounted dozen, you'll lose all the savings of your discount. Buy your hardware through catalogs rather than at retail hardware stores, as you'll get a better selection at a lower price. Many catalog dealers give a discount on bulk purchases. If they don't list a discount, or if the discount is for a number of pieces that doesn't suit your needs, give the people at the catalog a call and try to haggle a

> ## Handy Hint
>
> You need to create an accurate list of all the parts you'll need for a project. This means studying your design carefully and drawing out joinery and so on so that you know exactly how thick, wide, and long all your parts will be.

bit. They want to sell to people who are in the business because that way they move more volume with less handling.

You can buy your lumber and other supplies tax free if you plan to resell the finished product. To do so, you must register with your state to get a resale license. You then give your resale number to the lumberyard or other supplier, who sells you the goods tax free. This means that you must charge tax when you sell the piece (unless you sell it wholesale) and pay that tax to the state. Thus, the end consumer, not you, pays the tax. Having a resale number can get you discounts as well. Most lumberyards have a lower price schedule for "those in the trade" because they want repeat commercial business. Flash your resale license at them, and they'll treat you with respect (even though all you did to obtain it was fill out a form and send it to the state).

Keeping Track of Your Time

Keep time sheets on each project or group of projects you do. How you do this is up to you, but keeping some record is the only way to find out how many dollars per hour you're making on a given piece. You might keep a very detailed record showing time in and time out on a daily basis, or you might keep a general record showing how many days you spent on a certain project. However you do it, be sure to record in each entry what procedure(s) you did during the time period. After you finish a project or group of pieces, do a summary breakdown of the percentage of the overall time you spent on each operation. If a certain operation took far more time than you originally thought it would, you might need to redesign the project so that it will take less time to complete the next time or use a different procedure to make that part of the project more time efficient.

Ultimately, you should be working toward finding out how many dollars per hour you're making on your pieces once they have sold and you have a figure to use for your pricing formula. You'll likely find that on any given day you spend only about six hours working on your projects; the rest of the time is consumed by trips to the hardware store, telephone conversations with potential clients, working on improving your shop, and anything else that is not directly production related.

What do you do with all your time? At some point you simply must face the fact that all this time is the necessary overhead of running a business, just as it is in the executive offices of any factory. Executives might never turn a wrench, bore a hole, or sweep up sawdust, but their part of the operation is needed as well. You must include all this "office time" on your time sheets as well.

Pricing Formula

Your basic pricing formula is as follows:

$$\text{price} = \text{cost of time (hours} \times \$/\text{hour)} + \text{material costs} + \text{other expenses}$$

Your cost of time needs to include more than just the time spent cutting wood. It must include overhead time as well. You must set your hourly rate at a level at which you feel you're making an adequate profit, that is, making enough to make it worthwhile for you to do this at all.

Other expenses include rent, the cost of machines, the costs of selling (e.g., photography), and gallery or craft show fees. You must divide the cost of these expenses among each project as you see fit.

There's no easy way to accurately divide all these expenses into each project. Thus, most shops simply charge a flat "shop rate" that, from experience, they know will be enough to cover all the incidental expenses they encounter beyond project-specific material expenses. This makes pricing simple—you charge a flat shop rate per hour, plus materials.

Let's take a look at my jewelry box project to see how this all works out. I designed a small jewelry box for a birthday gift, then decided to make ten instead of one to take advantage of production time-saving techniques and sell the other nine. Because all the boxes have small parts, I was able to use lumber very efficiently. Each box took 4 board feet, so I bought 50 board feet so that I would have a bit extra to cut off splits and use as test pieces. At $2 per board foot for myrtle wood (a good deal from a local shop going out of business), I spent $100 on lumber, or $10 per box—very cheap.

Then I built the boxes, using finger joints on the box corners and for a shelf for each box. I did the finger joints on a table saw with a production setup, then glued them up, assembled the shelves, and finished them. Total time: 250 hours, or 25 hours per box.

The only hardware were hinges and pulls for the lid: 20 hinges and 10 pulls cost $80, or $8 per box. I finished them with wipe-on oil (which I already had) and used only half a can, so I didn't bother to add this cost.

I figured my time at $10 per hour (remember, this was 15 years ago). Using our pricing formula, here's how it adds up:

$$\$10 \times 25 \text{ hours} = \$250 + \$18 \text{ wood and hardware}$$
$$= \$268 \text{ per box}$$

If the gallery had accepted my boxes, it would have doubled the price to $540 per box (see the section "Pricing Realities" in this chapter). The gallery didn't accept them, but if it had I bet they would have asked me to lower the price, as in the late 1980s such a box just wouldn't sell at that price. I was able to sell many of them to friends for $200 each, reducing my hourly rate to a bit over $7.

GARY'S STORY

Wooden boat building has been the chosen path for Gary Blair of Santa Cruz, California. While in high school, his parents introduced him to a local boat builder who immediately took him on for the summer and then on weekends during the school year. He then branched out on his own researching boat designs and eventually building 34 boats, from 8 feet to 38 feet.

His research fascinated him, and over the years he began to study boats and seafaring in early times. Now he plans to write a book on the subject, covering issues such as early Mediterranean trade by boat as well as early water travel in other parts of the world. His current boat project, a 38-foot mahogany sail and power boat, will be named the *Minoan,* and Gary hopes to sail it to the Mediterranean to do further research.

Analysis

On reviewing my time, I saw that I spent about half the total time doing the finger joints and the shelves. Although finger joints are done on a production setup, it is still time consuming. People liked them, but the beautiful myrtle wood was the main selling point. Using simpler, less time consuming joinery would have reduced the time per box to under 20 hours, in which case I would have received the $10 per hour I was hoping for.

Here's another example: A friend of mine commissioned me to build a pine table with turned legs. I drew it out, totaled up the anticipated lumber costs and time, and gave her a figure, which she agreed to.

Building such a large boat alone may seem like a daunting task, particularly when the builder drives a city bus full time, but Gary is persistent.

"If you take a bigger project one small step at a time, it seems much simpler and easier," he says. "Don't let others discourage you, diminishing  your potential capabilities. You can figure out how to do almost anything if you really want to."

But then I went to the lumberyard and found that I had to buy more lumber than I had planned because the pine I needed came only in certain lengths that didn't work well for what I needed. I had to buy a lot of extra because I was going to have a lot of waste. Then turning the legs for the table took much longer than I thought it would because of the fancy pattern she asked for. Here's how the formula worked out, both as planned and in reality.

Planned formula:

40 hours × $10/hour = $400 + $80 wood = $480

Formula in reality:

50 hours × $10/hour = $500 + $120 wood = $620

My friend agreed to a final figure of $600 because she liked the table and understood when I explained to her the additional costs.

Pricing Realities

As the saying goes, something is worth what someone is willing to pay for it, and this is the stark reality that every manufacturer must face. In the final analysis, many people don't really care how long it took you to make something; rather, they care about whether they can get the same or similar item elsewhere at a lower price. They simply measure the desire for the item against the pain it causes to their bank account, and the subjective result is the presence or the absence of a sale. But in most woodworking you're faced with attempting to sell a luxury item that most people don't really need. Although your item is functional, it costs more because it is hand-made, and the customer is faced with the choice of paying more for that beauty and craftsmanship or buying an ugly but inexpensive plastic replacement.

Here's how the reality of pricing will work out. You'll walk into a gallery or shop with your item in hand (or a photo of it), and the gallery owner will compliment you on the piece, usually sincerely. If she decides that it's appropriate for her shop, she'll explain her terms. Most galleries and shops take 40 to 50 percent of the retail price of the item. If the item sells for $100 with a 50/50 split, the gallery gets $50 and you get $50. This might seem high, until you look at all the expenses required in maintaining a retail store with items that might not move quickly. Thus, you take your base price, then double it, and that's the retail price of the item. The gallery owner will tell you whether she thinks she can sell the item at that price. If she doesn't think she can sell it at that price, she'll ask you whether you'll take less. Now you have some hard thinking to do.

When you participate in a craft show, you'll have additional expenses that will increase the price of your items. These additional expenses are the show's entry fee and your transportation costs. However, you won't be paying a gallery owner's percentage at a craft show. Set your prices on the basis of what you need to make the event work for you, considering your additional expenses. Divide those expenses by the number of pieces you expect to sell, then add that price to the price of each piece (or proportionately if you have differently priced pieces). The good thing about craft shows is that they attract customers for the specific purpose of buying craft items, and some sellers at craft shows sell a lot of items in a short time.

If you're making a custom piece for a customer, it's best to come up with a flat fee at the beginning rather than asking them to pay you for time and materials. Time and materials is the best way for you, but if the time adds up to far more than the customer anticipated, you might have a big problem on your hands when he decides not to pay you. You need to plan carefully, figuring how much time you'll spend making a piece, then adding a little extra, as it always takes longer than you think it will. Present this number to the customer; if she balks, maybe you're better off spending your time weeding the tomato patch.

When you become an experienced woodworker and have developed a reputation and a recognizable style, you'll be able to tell people exactly how much you're willing to work for per hour. If you have a lot of work coming in, you can turn down work when people balk at the price. When you're starting out, you'll need to adjust your hourly rate to what the market will bear. But hang in there—every piece you sell is a calling card that tells people about a craftsperson who makes nice items that are worth seeking out!

Selling Your Woodworking

IN THIS CHAPTER, WE'LL TAKE a look at what it means to have a vision and a plan for selling your woodworking, where you can manifest your vision and plan, and interviews with people who have sold their work or who provide venues at which you might sell.

To sell your woodworking, you need to find the right group of people—those who want exactly what you have to offer. How you do this will depend on what it is that you have to offer. For example, a craft shop is not a good place to sell custom built-in cabinets because you can't show something on a gallery floor that hasn't been designed yet. Word of mouth and newspaper ads are the best ways to reach people who want you to come to their house with a photo portfolio and help them design a custom built-in. Conversely, word of mouth and newspaper ads won't work well for finding weekend tourists looking for knickknack carvings or small household items. Rather, these items will catch someone's eye sitting on a craft shop's shelf.

Let's refine this idea further. The leads you get from your ad in the newspaper and word of mouth might land you a potential client who wants something that is beyond your capabilities or beneath your desires to build. In this case, you have reached the wrong client, and you need to know when to say no. In addition, you should be careful about which type of shop you choose. Placing your work in a gallery or shop where it won't sell because the wrong people frequent that shop won't help you or the shop owner. Usually the owner knows best whether your work is right for her shop, but if she takes it and it doesn't sell, that should tell you that you might be better off showing your work elsewhere.

Having a Vision and a Plan

To find the right customers, you need to look closely at the kind of work you want to focus on and then form a plan for selling your work. The following examples show you how you might do this.

A carver takes an interest in using some local woods to make small figurines, which his family and friends like a lot. He has a vision to make many of these and to sell them. However, a local fine-arts gallery doesn't want them because they are "too folksy." The carver knows about a nationally advertised craft show in a large city several hundred miles away but realizes that his work is not of the same caliber as that shown there, so he doesn't bother to apply. With some investigation, he finds out that in a closer, smaller city there is a local craft, food, and music festival held in the summer. He applies for this festival, shows his work, and sells a fair number of carvings. At the festival he talks to another crafter who tells him about a craft shop in another town that sells local crafts as gifts to tourists. He makes a trip to that town, even though it is far away, and places his pieces, which sell fast enough that the owner asks for a steady supply of his work. Now the carver just mails off a few pieces now and then to the craft shop while getting ready for next summer's festival.

A hobbyist woodworker decides to get serious about furniture making, and after refining her skills she makes a beautiful table. She is determined to establish herself as a high-end furniture maker. A local craft gallery that sells mostly small items as gifts and knick-knacks accepts the table because the owner likes it and hopes to sell it. However, the table sits in the gallery for months and doesn't sell. She then does some research and finds that there are several high-end galleries that sell expensive, well-made furniture and art in a city half a day's drive away. She makes a trip there, looks at the galleries, and shows photos of her table to the owners. One agrees to take the table, which she then brings, and after two months it sells. Meanwhile, she has built another, which the owner agrees to take to replace the first. Later, the buyer of the first table calls her to ask whether she can design a matching table for their living room. She goes to their house, and they design a similar table, which she then builds for the customer.

A woodworker makes some small boxes, which a local craft shop accepts and sells. The owner tells the woodworker to make the boxes simpler and less expensively, which he does, and these sell well. However, over time he gets tired of making the same thing repeatedly for a minimal amount of money and wants to make more attractive boxes that sell in an upscale environment. He learns about several national craft fairs that sell high-end work and decides that he wants to make and sell work in those shows. He designs elaborate, beautiful boxes, then has some quality photos taken and applies to several of these craft shows and gets into one. He makes two dozen boxes, packs them in his truck, and drives off to the big city to sell his wares. They sell well, and even though after expenses he is not making a lot more than when he was making simple boxes, he feels that he is expressing himself more and finding more fulfillment.

These examples illustrate how you must focus on what you want to make and on where it is most appropriate for you to sell your work.

Selling Venues

Here is a careful look at the major sorts of venues where woodworking is sold. Don't be afraid to experiment with other more unusual ways of reaching customers. Just remember that you need to reach the right customers for your work.

Handy Hint

Always get a signed agreement stating the terms of sale so that there is no confusion later about who said what. Be sure that the agreement states when you will be paid, usually 15 to 30 days after the date of sale. If you sell a lot of small items, keep careful track of how many sell and when, and be sure to stay in regular contact with the shop owner about what sells, what doesn't, and what money is due when.

Galleries, Craft Shops, and Other Retail Outlets

A gallery or craft shop is a retail store of sorts. The owners have a vision of what they want their shop to be and what sort of work they want to be known for. Although some shops accept all different kinds of things, most have a particular focus and stick to it so that they will get repeat customers who know what to expect from the shop. This is why it's important to find a gallery or shop whose focus is at least close to the work you are doing and to avoid those whose focus is not.

There are a limited number of high-end furniture galleries, and most of these are in large cities. Craft shops that sell a broad range of arts and crafts will often take some furniture, but they prefer smaller items because they take less space and sell faster because they are generally less expensive. Some antiques shops will take new furniture if it is reasonably priced and similar in style to the things they have. Where there are tourists there are gift shops that sell most items at lower prices: $50 or so and some things over

$100. Craft malls are large buildings such as old warehouses or supermarkets that contain a large number of small shops full of a wide range of items. Each shop might be an independent business, or all the shops might be run by one person. These businesses attract a lot of people because of the quantity of items for sale.

In some craft shops and galleries, the owners will buy your work from you outright, but in most cases they take it on consignment, meaning that they pay you after it sells. You sign an agreement stipulating the terms, which often state that you get a percentage of the retail sale price of the piece (50 percent to 60 percent is common) and the gallery gets the rest. Then they put a price on it and try to sell the item at that price. If it doesn't sell, they lower the price and wait and see. This is where you find out what the market will bear.

For higher-priced items, they'll simply ask you how much you want for it, then you sign an agreement that states that they will pay you that much when it sells, and they put whatever retail price on it they want.

Shows

In the same way that craft shops and galleries come in all shapes and sizes, craft shows vary widely in scope and quality. The Smithsonian Craft Show is considered by many to be the premier show in the country, although others rival it. This show features a broad range of crafts and gets more than ten times as many applicants as it has booth spaces, and people from all around the country apply. Check out its Web site (see the resource section) to view samples of some of the finest crafts available today.

On the other end of things, you'll find local arts and crafts festivals often combined with music, wine tasting, county fairs, or other local events. These events might make an effort to find local craftspeople like you, but sometimes they emphasize sellers with imported or mass-produced goods. This changes the character of things and might not be quite what you want. In addition, be aware that in some parts of the country so many summertime craft shows take place that many people make a full-time job of selling at these, going from one town to the next on the circuit. If you're a part-timer, you likely won't want to sell items that these commercial craft show retailers can sell far more cheaply than you can.

In the middle are regional craft shows, which emphasize a variety of crafts, including different kinds of woodworking. These are sometimes produced by entrepreneurs who do this as a living (and a passion), whereas at other times they are produced by craft organizations, such as the American Craft Association or local craft guilds. These shows will be easier to get into than the Smithsonian Craft Show, and you won't have to go as far to participate.

Entry into craft shows is done by application and usually includes showing slides of your work. For local craft shows, the application deadline is anywhere from a few weeks to a few months from the show date. You can find out about such shows by talking with the local chamber of commerce or local craft galleries and craftspeople. Applications for larger craft shows that are regional or national in scope are due six months to a year before the show, so this takes some planning on your part.

For all craft shows, you fill out an application and send in about $25 for a processing fee as well as 35-mm color slides. Their entire decision about who to permit in the show is based on these slides. The jury doesn't have the time or the resources to read resumes, letters of recommendation, or published articles about your work. A jury of several art and craft professionals sit and look at all the slides and read your brief description. This is why it is critical that you get good photos of your work (see chapter 10). Jury members might have to look at 200 slides in a single day and then make the

Handy Hint

You can find listings of upcoming shows, application due dates, and addresses in *American Craft* magazine and *The Crafts Report* (see the resource section).

best decision they can, so a lot of their judgment boils down to a basic subjective response. Remember that good photos result in favorable subjective responses, whereas poor photos don't.

Although some people believe that this sort of selection process is limited, it really is the only way it can be done. In some ways it's not fair because someone who simply can't afford to have quality photos taken doesn't have as good a chance. The jury is simply trying to find what they feel is most appropriate for their show and what will most satisfy buyers so that they come again next year and the show gets favorable reviews.

When your application gets accepted, you will be expected to pay a hefty booth fee, anywhere from $500 to $1,000 or more. This is the only major fee that you pay to the craft show. It's impossible for them to charge you a percentage of your sales because they have no way of verifying all your sales. A booth fee makes it easy. However, you need to decide whether you'll sell enough work to justify the cost of the booth fee. Of course, you can charge the full retail price for your work, unlike in a craft shop or a gallery, and you can sell a lot of work at a craft show because people come there to buy.

Other costs of attending a craft show are transportation to and from the show and hotel fees and other living expenses while you're there. Some people who sell at a lot of craft shows have motorhomes wherein they can transport their work, sleep, and cook to reduce expenses. Although most shows take place over only two or three days, you can rack huge hotel and restaurant bills if you're not careful.

You'll need to have a booth setup to take to the show, and some shows will ask for a photo of your booth in the application process. They want to know that your booth will look good and be a positive contribution to the show. Some shows will supply pipe and drape, which is light piping set up around your booth space with cloth drape hanging from it to separate you from your neighbors. You might be required to supply your own pipe and drape. In addition, you'll need other things, such as tables, display racks, chairs for you and customers, and anything else to display your work in its best light. People who sell at a lot of craft shows often have a carefully made setup that packs up easily into crates for shipping. Some booths are very elaborately done, transforming a 10-by-10-foot space into a small but personal gallery space.

Questions to Ask Craft Show Producers Before You Go

Attending a craft show is an expensive and laborious proposition, and you have the right to know at the outset what is involved. Don't be afraid to ask a few hard questions of the producers after you've been accepted and before you take the plunge and write that booth fee check. Here are some things to ask:

■ How much advertising will the producers do for the show? (How well the show is attended by buyers depends on how many hear about it.)

■ What was last year's attendance?

■ Do they have figures available on door ticket receipts and exhibitor sales?

■ Will they make any special facilities available to you that you might need, such as an electrical connection or an unusually shaped booth?

■ Can they show you a layout of all the booths, including where yours is, so that you know that you aren't located in a corner that will get less traffic?

■ What sort of scheduling will be required for setting up and taking down your booth?

■ Will there be help available, such as helpers setting up and taking down heavy objects and booth sitters who will watch your booth during the show so you can take a break?

Selling on the Internet

Probably the best piece of advice you can get as you begin to re-search selling on the Internet is to beware of the hype. You are not going to get rich overnight selling your woodworking on the Inter-net, but it does present a unique way to reach customers that other-wise would not be aware of your work. At no expense, viewers can pull up photos and descriptions of what you're making and order items without even licking an envelope. But would you buy a piece of woodworking by mail order or over the Internet after seeing only a photo of it? If you're making large, expensive furniture, your chances of selling on the Internet are probably less, although your photos might attract potential clients in your area looking for cus-tom furniture. Easier to sell will be smaller, inexpensive items, which are mostly what you find on Web sites that sell crafts.

The Internet as a commercial medium is still developing, and many different opportunities likely will arise through time. Let's look at some of the main ways in which people get pictures of their work onto other people's computer screens.

You can create your own Web site by buying a program that gives you the tools to do so or by learning HTML (Hypertext Markup Language), the basic language of instructions that all Web sites are built on. You build the site in your computer and then up-load your files to an ISP (Internet service provider), which is the server computer that connects your computer to the Internet. The files that constitute your Web site reside in the memory banks of the ISP's server, so the people browsing the site see the material through the server, not through your computer. For people to order your work, your Web site (and your business) needs to be set up to take credit card numbers. You need an ISP that has a secure server so that this can be done safely. In addition, the ISP will have special programs that process the credit-card transactions.

Creating your own Web site takes a lot of time. Then the question is, how many people are going to find it? You can set up the name of your site such that when people enter key words such as "woodworking" or "crafts" into a search engine, they will find your site listed. Even better is trading links with other Web sites. A link is simply a button or specially coded string of text on one site that takes you to another site. If you and several other woodworkers who sell on the Internet trade links, then people who happen to find one site can, if they choose, go to the other sites—and they are likely to go there because they see that the linked site concerns the same subject.

> ## Handy Hint
>
> You can hire someone to build your Web site for you. Depending on how large or complex you want your site to be, the cost of doing this can range from a few hundred to thousands of dollars. It's best to start small to get your feet wet and then decide where to go from there.

Another approach is to have your own listing or a Web page on another Web site. Many such "host" Web sites exist that do a variety of things (see the resource section). They might simply show a photo of your work and provide an address, or they might give you a page or two (that you can design yourself, using their instructions) to show your work. They can also provide a way to sell your work with credit-card orders. Many online craft malls take this kind of approach. Some of these are based in actual stores where some or all of the crafts are located, whereas others exist only in cyberspace and each of the shops where the work is produced.

Some host Web sites are free, and some charge a monthly fee. Before you pay to have your work put on a site, ask how many people visit the site, but be wary if this is a very high number. Ask whether they have any actual sales figures available. Contact a few of the other sellers on the site and see whether they have had satisfactory results.

A variety of sites on the Internet provide information to craftspeople on subjects such as making your own site, selling on the Internet, and other craft-related issues, from techniques to business information. These sites have links that will take you to similar sites. You can surf the Internet for hours browsing such sites and finding a lot of useful information. Some of these sites are listed in the resource section.

Other Ways to Sell Your Woodworking

In addition to those already mentioned, there are a number of other ways to market your wood products. Some of the following suggestions will involve an actual sale of your wares, while others will serve more as a way of exposing your skills and samples to a large audience.

Guilds and Cooperatives

Sometimes a local crafts guild or cooperative will set up a craft shop or gallery, in which case you help run the shop yourself. Be sure that your commitment in time is clearly spelled out when you do this so that you avoid misunderstandings later.

Designers and Interior Decorators

Many people trust these professionals to find furniture and other things to grace their living space. Usually they'll be happy to know about what you do so that they can offer their clients a broad range of possibilities. Give them photos of your work and let them know that you're adaptable.

Charity and Nonprofit Auctions

Often these kinds of organizations take donations for fund-raising auctions. Your donation gives you good exposure, which can lead to new customers.

Open-House Studio Tours

Local craft or other organizations sometimes set up tours of local crafters' studios so that the public can see how the craft is made. If you can get into one of these, people will literally come to your door to see your work.

In the Trenches

Now let's talk with some people who have actually been out there in the world of making and selling woodworking and other crafts. We'll see what a professional cabinetmaker has to say about the side jobs he does out of his garage on weekends as well as a couple who carve and paint unique statues and figurines. A full-time post office employee tells us about his part-time ornamental turning career. Then we'll talk with the owners of a gallery in Santa Cruz, California, as well as the producers of a craft show near Los Angeles.

Brian Argabrite: Moonlighter Extraordinaire

Brian Argabrite is a professional cabinetmaker in Santa Cruz, California, who works in a small shop that produces custom cabinets, furniture, and whatever other business the owner is able to find. This kind of work is a bit beyond the reach of someone who is just beginning in the craft, but Brian also maintains a separate, part-time business doing occasional things in his garage on weekends and sometimes on weeknights. Doing so, he has learned firsthand the ins and outs of developing a small, part-time career by word of mouth.

His shop is a testimony to what you can do in a small space. It's a one-car garage with a small overhang extension on the front that contains everything he uses. The table saw is smack in the center of his shop, giving him lots of room to rip sheets of plywood if he has to. The extension tables on the saw function as workbench, and surrounding the saw are other tools, such as a band saw, a drill press, and a compound miter chop saw. All these tools are small enough that they can be easily moved about as needed. In the corner is a huge 90-year-old cast-iron planer, a bit larger than perhaps he really needs, but Brian fell in love with the machine when he first saw it and had to have it.

How do you get most of your work?

"It's 99 percent word of mouth. The bottom line is, if you do good work, and you do it with a passion, people will appreciate what you are doing and hire you. You just have to open up to it and say 'I can do your cabinets' and show some skill. In a weird cosmic sense, if you put out the energy in a certain direction, it'll work, it'll happen.

"My first break came when a friend of mine referred me to a man looking for workstation cabinets in a hair salon. I took this job just for the experience—I wasn't expecting to make much from it. I did an inordinate amount of time with design and looking at all the details that first time around to make sure I did it right and the customer was satisfied. A year after the salon opened, it was voted best hair salon in the county, which felt good.

"The customer was looking for my input on the design on this job. I had a lot of liberty with it. He was a bit nervous at first because he knew I was inexperienced. He came to my house to see pictures and examples of what I had done in my own house. He was very anxious because he had to have the store open by a certain time and didn't know if I could produce. I must have worked straight through about 40 hours at the end of that job to deliver and install on time. But it was worth it because he was satisfied.

"As a woodworker selling for the first time, you gotta be straight up. Be honest, look people square in the eye, and be on time, or tell them if you have to be late. On that first-time interview you gotta be on time—don't make that mistake! As a beginner, you can't necessarily be as forceful, as much in control, as when you've been at it for ten years and have a lot of work coming in. But you know, some of these experienced guys have a real chip on their shoulder, and they let the customer know it, like they should be grateful or something. I don't want any part of that."

What happened after that first job?

"The hair salon was a business that sees a lot of traffic, lots of people see it, and I got more work from that. But first I did another job for the same client, a whole small grocery store full of simple shelves and counters. It made a difference to him to have me because he needed to convince his financial backers that the store should be re-done. He told them that he had a trusted cabinetmaker, me, so he was able to convince them, and then there was no competition for the job—I had the job.

"After that I did some work for a clothing store, a referral from the owner of the hair salon, and then some outdoor furniture for the grocery store. Once I got started, it just kinda kept coming."

What sorts of things did you learn once you got started?

"Once I got started, my experience told me I had to be flexible with changes. It's critical to show the ability to adapt when the customer changes their mind. This isn't unusual—it's usual in the custom wood business. You have to expect changes, don't be alarmed. It's a big mistake to overreact. They are coming to you because they want satisfaction, and that means accommodating their changes. You just can't take a project and assume it's set in stone.

"But it's real important to make sure you get paid for those changes. You can't just say, 'Well that's more work' because they will say, 'Okay, do it,' thinking that you will absorb the cost as part of the contract. On my first job I made sure that I told him right away that a change was going to cost more, and I let him know how much as soon as possible."

How do you put a price on your work?

"A flat bid for the whole job is best if you know what you are doing, otherwise you go by honesty and just charge the customer what it finally costs in terms of total time and materials. When I bid a job, I carefully look at what it will take beforehand. I've got legal-size yellow pads full of sketches, notes, and tallies of numbers on hours and materials for each of my jobs. I give them a bid at the outset, then I charge 50 percent down and 50 percent when the job is finished. Sometimes, if I have the materials and like the customer, I tell them not to pay me at all until it is done. That way, it is a trust-based system. They leave you alone, you work. I never have been taken for payment that way.

"Take your cost of materials, how much time your think it will take to do the job, then take the hourly rate you want and add $5 an hour. That $5 an hour is for overhead. You are spending X amount per hour just standing there for machine costs, electricity, etc. Then, as you do the job, keep close track of your time. It's important to me to know how much time it took to actually do the job from design to completion, and with simple math, how much I made an hour. Then I use this information to help me come up with more accurate bids on future jobs.

"Also, as I do the job, I keep track of all phone conversations I have with the client. On this date, this was said, on this date, this was said. This helps me total up all the changes that occur."

What kind of publicity do you use?

"I ran a deck and fence business, did lots of publicity like fliers, classified ads in the small-business section, coupon mailers (which got me nothing), but by far the best was word of mouth. The ad in the small-business section worked fairly well. I left it in for a few months, then took it out because I then had lots of work. It's important to keep up contacts, too. A former boss has let me use his machines from time to time when I have needed them, and he has referred me to a few people."

What advice on tools can you give people?

"You have to have good tools—to make money you can't be using garbage. I get any tool I come across that's a good deal. I have seven routers. If I found another, I'd get it. I had three of them set up once, then used a fourth for something else so I didn't have to break down a setup and set it up again. Saved me time."

What's most important to you about running your part-time moonlight business?

"What's important to me is that I produce quality work, and I tell the customer that. That's not just so I get a good reputation, it's so I can sleep at night. If you do good work, you'll get more work, that's true, but it's just important to me to know that what I do is well made.

"I like the advantages of being small. You can control your time, do a project, then don't do any for a while. Sometimes it's nice to just be left alone for a while before I take on the next project. There's a jump you go through to go to a larger business with employees and all that. Once you are there, you have to keep a steady stream of work coming through to pay all the bills. You can't just send everybody home for a few weeks if there's no work. With a one-person operation, you can quit for a while and enjoy life, you know?"

Carol Harper and Dan Adair: Makers of Painted Carvings

Recently I attended a local art and wine festival in Boulder Creek, California, to see what other craftspeople were doing. There I found Carol Harper and Dan Adair, with a booth set up showing the painted carvings that the two collaborate on. The wood used is all storm downfall from local forests. Dan and Carol are computer buffs and met through the Internet, so it's no surprise to find that they sell their work partly on the Web. You can visit their site (www .sasquatch.com/ ~ mars/INDEX/Index.html) to see photos of their work, which depicts numerous characters from mythology, both ancient and modern.

> ## Handy Hint
> Be sure to try many different types of woodworking projects. That way you will find not only which ones you enjoy the most, but which ones you are best at.

How did you get started doing carving and painting?

"We both carved and painted independently for years. When we fell in love, we started doing it as a joint product. I had never carved anything specifically for painting, and Carol had never painted a carving. As we've continued to work together, we're cooking up our own techniques to deal with the new medium and—hopefully—getting better all of the time."

How did you get started selling your painted carvings?

"We started by walking into a local art gallery and talking to the person who was gallery sitting. He encouraged us to bring our work in and enter it in the next show. The response we got was so positive that we realized we had a salable product."

What are all the different outlets in which you sell your carvings?

"We sell through the local art gallery (Santa Cruz Mountains Art Center, Ben Lomond), through craft shows, we have a couple of orders from a toy store, and we have sold a couple of pieces through our Web site."

Tell me specifically about the Boulder Creek Art and Wine Festival—how it worked out for you and other crafters.

"It worked out both well and poorly. The 'poorly' part was from the point of view of sales. I've talked with nearly all of the local artists and crafters who took part, and they all report extremely poor sales. They all attribute this to the fact that the promoter of the show had included massive numbers of resale and import booths. This drew a crowd which was only marginally interested in art and had no idea of the work involved in crafting.

"It worked out well from a point of view of networking. We met several people who are interested in our work and learned a great deal from talking with other crafters and artists."

What advice can you give beginner woodworkers about how to sell their work?

"Network, network, network. In the words of an old redneck friend of mine from Texas, 'You got to have a lot of poles out to catch a fish these days.' (1) Find the nearest *small* town near you and join the local art guild or gallery. Not only do they tend to be very supportive of new crafters and artists, but they are also an invaluable resource for information about shows. (2) Participate in shows and don't be discouraged. Look on them as research initially rather than a moneymaking endeavor. You're there to look at other people's work and prices and to meet people who can help you. Crafters and artists *like* to help other crafters and artists, so don't be shy about asking questions. (3) Contact small shops and see if they're interested in carrying your pieces. Be willing to lend them on consignment because crafter's shops tend to run on a shoestring. (4) If you've got an Internet account, get a Web page up. Participate in the crafts and artists discussion groups: they are a wealth of information about techniques, shows, and prices."

How do you price your carvings? Do you have a formula, or is it an educated guess?

"We researched extensively over the Web to see what other crafters were charging for similar pieces. We charge more for the pieces that we sell through the galleries because they are taking a 25 percent commission out of the sales. You also have to figure in a consignment fee if you're going to sell through small stores. It's very tricky getting the money that you need out of the piece without overpricing your work."

How do you advertise or otherwise attempt to find customers?

"Our main source of advertising is over the Web. Once I set the pages up, I joined a number of Web rings. These essentially hook you into other artists' and crafters' sites so that you have a chance of several 'hits' a day over your Web site."

Jon Sauer: Ornamental Turner

After a full day as an executive with the U.S. Postal Service, Jon Sauer comes home to his wife and kids in Daly City, California. When he finds some free time, he goes into his garage and, in a corner less than half the size of the two-car garage, works with his Holzapfel lathes. Over 100 years old, these machines are still serviceable in the able hands of this man, who has been

Handy Hint

Ask around at crafts and woodworking shops to find out which types of items are selling the most. Most shopowners will not have a problem providing such information.

using them for years to make such things as small lidded boxes, vases, and other little containers. What started as an intriguing hobby is now also a well-established business, with Sauer selling his pieces in numerous galleries and some of the best craft shows.

How did you get started selling your turnings?

"I started selling my turning by word of mouth to my friends and their friends. I then walked into a gallery and showed them some of my pieces. They told me to leave their gallery and try down the block. So I did just that and tried the flea market.

"My work was too good for that and did not sell well there. I then went north to collect wood on the beaches and stopped in at a gallery in Mendocino and told them my story of not being able to sell my work at the flea market. They asked to see some of my work, which I had in the car, and they actually wanted to keep it to sell in their gallery on consignment. Sounded good to me, so I left it all with them, about 14 pieces. They all sold within a few months, so I had to make more for them. I have been in this gallery now for about 20 years. I soon went into other galleries and did the same."

Where do you sell your pieces?

"I now sell my work through galleries across the nation and even a few abroad and do a few high-end craft shows a year (American Craft Council Craft Shows, Smithsonian Craft Show, and Contemporary Craft Shows)."

What advice do you have for people trying to get into craft shows?

"To get accepted into a high-end show, you need good photos of your work. Pay the price to have a professional take quality photos of your work. Always send originals in for the selection committee. They look much, much better than duplicates. If you have good work, and if it's what they want and are looking for, it will be accepted."

What advice do you have for people trying to get into galleries?

"Same as the previous question, but shop for the gallery that is on your level. Also, most of the galleries will not purchase your work, so be prepared to give it to them on consignment. Make sure they pay you monthly for any work sold from the month prior."

How should craftspeople price their work?

"Pricing is very difficult. Do not price your work too high, as it all should be. Keep it affordable, as when it sells you can now sell the pieces that you have just made. This keeps you working, and as time goes by, your work gets better and better, and then the prices go up and up. Then you can ask for the big bucks, and you have the name to command it. Be patient, as this might take years."

What are some good ways to advertise?

"Send photos of your work to magazines that print work of artists. I have postcards made of my work and send them out to collectors, gallery owners, and the general public regularly."

Can you suggest any other resources?

"Join your local woodworking organization and any national groups that are of the same interests as yours. Subscribe to magazines and periodicals to keep you informed of what's happening in today's ever-changing world. Some list opportunities to enter shows, fairs, and special museum shows."

What general advice do you have for newcomers?

"Take one step at a time. Remember first we could only crawl, then we walked, and then we ran. Be patient, and soon you might fly."

Artisan's Craft Gallery in Santa Cruz, California

Juleen Lisher and John Lisher have been operating Artisan's Gallery in downtown Santa Cruz, California, for many years now. The gallery was started by a cooperative of craftspeople, but when they found it difficult to make ends meet, John and Juleen took it over. The gallery is located on an outdoor mall where many tourists and locals browse the many shops. Most of the items in the gallery are small items priced under $100, although they do keep a rocking chair on the floor by a certain local woodworker. That piece in particular has sold well through the years, but in general they shy away from larger pieces like that because they take up so much space.

Among the wood items they have on display are numerous small boxes, handheld mirrors, clocks, pens, serving trays, chopsticks, picture frames, clipboards, wine racks, spoons, forks, spatulas, salad tongs, and salt and pepper shakers. Items that include both wood and metal are money clips, key chains, pillboxes, and business card holders.

Juleen told me that a key factor in the success of woodworkers who sell in her shop is using production methods to reduce costs. Most of the items on display easily lend themselves to production runs, but the large cherry wooden spoons that I saw did not. Items like this still must be made one at a time, and as a result the price is significantly higher than the cheap ones you see in the grocery store. However, John told me, these cherry spoons look so much nicer that they sell well.

Did you know???

As a professional woodworker, you may be entitled to a certain number of tax deductions. You might be allowed to deduct for such things as equipment and tools, business costs, and training. It's definitely worth checking into.

What advice do you have for newcomers to selling crafts in galleries?

"Keep in mind that galleries and all businesses have to sell in order to pay the rent. Be realistic in setting prices."

How should craftspeople set the prices on their work?

"Objectively ask yourself, 'If I saw this piece in a shop or gallery, what would I expect to pay for it? If I made this piece again (or a third, fourth, or fifth time), how much time could I save, how much material could I save? Would the end product cost a lot less to make after the fifth time? Do I want to make five? Is selling this piece optional to me?' If you are making heirloom pieces that you are happy to keep in your family, then they are priceless, but if your expectations are to make something to sell, then you must keep in mind the buyer's willingness to pay for the product."

How best should someone approach a gallery?

"Call and ask for the owner. Say you are a woodworker and would like to show the owner your work for the purpose of putting your work in the gallery. I have looked at things in the backs of trucks, and if I am in, I will always try to see the work, but it is better to call first, then bring pictures or product."

Describe your selection criteria.

"I look for good design, good detail, good finish, and a salable price."

Describe your clientele.

"A clientele that is very appreciative of wood. Because our space is limited for furniture, we carry one handmade wood and leather rocker, but our other work is boxes, chopsticks, cutting boards, and wooden kitchen utensils."

Do you have suggestions for how people can advertise?

"Local craft fairs and exhibits."

What sells best in galleries such as yours?

"Functional work that people can use or give as gifts."

What should someone expect once their piece is placed in a gallery?

"We place work for a period of time, 30 or 60 days. Customer feedback is crucial: What are the customers saying, and what is their reaction to the work?"

What resources do you suggest for craftspeople?

"*Craft Report* magazine is good for suggestions on marketing crafts. Wendy Rosen wholesale shows always have a lot of woodworkers. Her shows are in Philadelphia in February and July. *Fine Woodworking* and *American Woodworker* are good magazines about the craft of woodworking."

Nancy Peck and Joanne MacDonald: Producers of the Art Furnishings Show in Santa Monica, California

For years, Nancy Peck and Joanne MacDonald had been producing various events for both profit and nonprofit groups. They decided to take these skills and combine them with their love of crafts to produce their own show. Thus was born the Art Furnishings Show, which had its debut in the spring of 1999 and is planned for coming years. The show's theme, furniture and furnishings, features a variety of furniture styles as well as wood sculptures, bookcases, beds, playhouses, and decorative items, such as mirrors, frames, clocks, music stands, and Judaica (menorahs). This is a fine craft show, attracting customers who are willing to spend more money for items of high quality. Exhibitors come from all over California and some from out of state.

What general advice can you offer newcomers to craft shows?

"Craft shows are one of the main venues that woodworkers use to show and sell their work; however, it is important to opt for the right ones for your type of work. By all means ask questions of the producers and of exhibitors who have participated in the past. If you are new to doing shows, admit it. You'll get more help that way. Most producers expect to introduce new talent to the public.

"In general terms, consumer shows are classified as traditional country craft or contemporary fine craft. How would you classify your work? It may not be a good idea for you to try to sell your finely crafted, large pieces of furniture at a street fair where jewelry, wearable art, toys, and wind chimes predominate.

"When it comes to planning your graphic design, deciding which inventory to show, or designing your booth space, always keep in mind the type of persons who will ultimately be attending the shows and respond to your work. By 'type of person' we mean the general tastes, geographic profile, socioeconomic status, and their experience attending shows and collecting.

"Don't wear yourself out haphazardly trying to satisfy the taste levels of everyone. Create your own original style you've not seen elsewhere—fine-tune and focus on it. Perhaps a unique aspect of one piece could be subtly incorporated in your other pieces. This will help to make your work identifiable and contribute to a visually attractive and cohesive display.

> ## Handy Hint
>
> "Try to match the value of your items with the audience. For instance, for casual festival-type civic venues, be prepared to offer some less expensive items, such as small boxes, birdhouses, window boxes, garden ornaments, or simple toys."
>
> —Nancy Peck and Joanne MacDonald

"Know that even the most successful veteran artists on the show circuit have unexplained bad stretches and unexpected triumphs. Some may do well at a particular show one year and not as well the next. So many uncontrollable factors contribute to attendance and buying moods, and there may be some competition within the show. One needs to look at his or her proceeds collectively over several years. Don't make major marketing upheavals or broad assumptions about the quality of your work based on your sales at one show."

Describe your process of selecting exhibitors. What are you looking for when viewing slides?

"Those artists who are accepted into the show have submitted photographs representing products that are unique, have a strong design element, have a high level of creativity and consistent workmanship, and finally have customer appeal relevant to the anticipated market.

"If you are serious about getting into the best shows, then spend the money to make sure your slides are shot in a professional manner. Poor lighting, poor backdrop, and poor product arrangements are factors that will compromise your chance of being accepted."

What sorts of woodworking will do best at craft shows?

"This all depends on the generalized preferences of a particular geographic area, socioeconomics, etc. Try to match the level of traditional style with the tastes of the attendees. Again, the profile of the attendees may dictate whether you should offer sculptural, functional, decorative pieces, small, large, or adult versus child appeal. When it comes to *craft* shows, the audience is generally interested in seeing new and unique pieces, finely crafted products, which they can easily incorporate in their daily lives."

What should one think about when preparing a booth?

"The requirements for indoor versus outdoor shows are different. A consideration is the ease with which your booth can be assembled. For most outdoor shows, you will have just a few hours in the early morning to set up, and some outdoor shows require dismantling every night of a weekend show. Indoor shows generally allow for more setup time. Be aware that you will be tired at the end of a show, and you will want to dismantle without too much effort or time.

"For outdoor shows, you will need to consider shelter structures, in the event of precipitation and wind. For indoor shows, invest in spotlighting to showcase your work. For woodworkers especially, soft, warm lighting can make an ordinary piece of furniture look extraordinary.

"Use open space to set off your products and make them more accessible. Don't overload your booth with everything you've ever made. Don't overwhelm the customer with a busy array of various products. Sometimes showing less of your inventory makes each piece seem more special.

"Think in terms of 'vignette.' It helps the customer picture your product in their environment. Birdhouses swinging from a tree branch will be far more delightful to view and be more effective than arranged linearly on tables.

"Many exhibitors do not make use of their wall space. Furniture is meant to be seen below eye level, but smaller pieces could be displayed at eye level. Make use of your wall space with blowups of your work or display shelving at varying heights."

What should one expect will happen at a show?

"Expect that people will want to negotiate your price or get a deal at closing time. Whether you succumb is up to you. Be prepared to take checks and cash, and nowadays most people like to pay with credit cards. Invest in credit card service if you plan to do many shows. Allow time sometime after the show for delivery of your large pieces. Expect sales after the show. Expect requests for custom work. Expect galleries or specialty store buyers to make you offers. Get educated about business ethics in those fields. Ask your peers.

"You have worked hard to prepare for the show, spent money, and have invested a lot in the emotion category. (That goes for producers, too.) You are venturing into unfamiliar territory. Everything may not go exactly as you had envisioned. Be prepared to be disappointed in at least one way, but try to maintain a positive attitude while the show is going on. Be prepared to meet new people, and you might even make new friends."

How can craftspeople advertise their work?

"We have found the most successful marketers are those who know that just one avenue will not suffice. The compound effect of all your marketing methods over the years will lead to ultimate success and minimal discouragement. This requires a long-term commitment. Few of us reach success overnight. Be tolerant with your work and yourself. Give yourself years before even considering giving up, and allow your marketing efforts to take root and develop.

"Other than consumer shows, here are other ways to advertise (and sell) your work:

- Color postcards, to be used at your booth and as direct mail to past customers (Also at your booth, show cards advertising your future gigs.)
- Pamphlets
- Newspaper publicity
- Trade magazines
- Web site

Handy Hint

Never stop learning. Check your local college or craft shop for classes on both learning your craft and being in business.

- Guilds and co-ops
- Galleries and specialty shops
- Interior decorators and architects
- Trade shows
- Silent auctions and fund-raisers
- Small boutique-type shows
- Open-house studio tours
- Store window displays
- Word of mouth

"Develop relationships with your competitors. You never know when they might be too busy to fill an order. The more successful artisans spend half of their time on marketing!"

chapter 10

Marketing Your Woodworking

HOW DO YOU GET OUT the word that you've made something that's worth finding? The trick is to reach the right group of people for what you're offering. You don't want to spend money or time on advertising that won't reach potential customers. For example, you could put an ad in your local newspaper in the small-business section. You might do well with an ad like this if you do custom cabinet work but not if you make small knickknacks, such as household items or simple furniture. You're better off selling these in a gallery or at a craft show.

However you advertise or publicize, you'll help yourself tremendously by having high-quality photography of your work. Making contact with potential customers really boils down to this: People want to know what you have made, and words don't tell them this as well as pictures do. Later in this chapter we'll look at how to get those good photos, but first let's look at some avenues in which you can use them.

Where to Advertise and How to Publicize

Word of Mouth

When other people do your advertising for you, you're one step ahead of the game. A good recommendation from a satisfied customer is worth its weight in gold. I've known woodworkers who bent over backward and even took a loss on certain jobs just because they knew that the recommendations they would receive from a certain client would greatly help them. However, this kind of advertising comes only after you have made some items for customers, so you'll need to seek another avenue of advertising if you're just starting. But when you do get that phone call from a stranger who has heard of you from someone else and who is looking for something you have made, you need to be ready to show that person a good photo portfolio to make the sale.

Networking

You'll often run into other woodworkers and people associated with the trade, be they clerks at lumberyards and hardware stores or tool repair people. Such contacts can clue you in about potential work and possibly recommend you to others, just as you can do for them. Again, be ready with that photo portfolio.

Newspaper Classifieds

Most local newspapers have a section in the classifieds for small local businesses to list their service. Proclaim yourself a custom cabinetmaker and furniture builder in your own small ad, and chances are you will get calls. But you need to be ready for those calls with a photo portfolio and confident advice for the potential customer about how best to approach the desired project and its cost.

The Internet

Some local Internet service providers (the company that connects you to the Internet) offer classified listings to their customers. These are like the newspaper classifieds, but they reside on your computer. Such a local listing might be of little use to someone in Germany looking for a woodworker to do a custom job, but this is a great way to reach potential local customers.

As discussed in chapter 9, many galleries and craft malls are on the Internet, and if you plan to present your work there, you'll need—you guessed it—good photos!

Postcards

Services are available that can provide craftspeople with color post-cards of their work (see the resource section). You send in a photo, and for around $100 you can have 500 postcards made. You can send postcards to galleries, craft malls, past customers, and any contacts who show any interest. This is the least expensive way to have lots of photos made; prints made by your local photo store will cost much more.

Brochures

A brochure is a glossy full-color page that includes photos as well as text describing you and your work. You can have the brochure folded so that it opens to reveal its contents, or you can leave it as a flat sheet, like a flyer. You'll need someone to do the brochure's lay-out, and this kind of service has become more common with the ad-vent of computers and page layout programs. Look in the yellow pages under "Graphic Designers" and "Printers" to find services that can help you design and print a brochure.

Having a brochure available whenever and wherever you sell is the best way to remind interested people later about your work. A brochure is easy to pick up and put away, then later they'll find it among their things and be reminded of you.

If you have a computer, you can design your own brochure fairly easily. Many inexpensive programs are available to help you do this. With a high-quality printer, you can print them yourself, or you can take a floppy disk to a local service to have them printed. If you design your own brochure electronically, you'll need to scan your photos (scanners have gradually come down in price), or you can have photos scanned to disk by professionals for a fee.

Business Cards

Having business cards printed is very inexpensive. Although you might already have cards for your other profession, have new ones made for your woodworking business, perhaps a set that says "John Doe, Woodworker." This lets people know that you're serious about your craft. Their confidence in you will depend partly on your confidence in yourself. Your local copy service might make business cards for you, or you can hire a graphic designer.

Press Releases

Your local newspapers and magazines are always on the lookout for talented local people who will make a good article or event listing. The press release is a simple way for you to say, "Hey, I'm here!" Editors get a lot of these, which they often sift through quickly and skeptically, but if you are persistent at sending them out, there's a

Handy Hint

Hiring a graphic designer to create your promotional materials can be expensive. If you live near a college that has students learning these skills, you might hire one to do it cheaper as one of their class projects.

good chance you'll get some attention. A press release should contain some kind of timely information, such as your introduction of a new product, the opening of a show featuring your work at a local craft shop or gallery, or a major commission you just received. Write up a brief, concise, one-page release that tells what you are about and the event that you are bringing to the editor's attention, and be sure to include contact information and a good photo.

Mailing Lists

Keep a list of addresses of interested people and customers. At a craft show or other public event, keep a sign-in book. Later, when you have a new postcard made of your most recent work, send the postcard out to everyone on that list. Send a thank-you note to someone who buys from you and include a big, hearty, sincere "Thank you!"

Artist's Statements and Bio Sheets

Potential customers often want to know something about you and your philosophy. Craft shops and galleries often will ask you for an artist's statement and/or a bio sheet for this purpose. To some people, the philosophy behind the craft is as important as the craft object itself. Some artist's statements are full of high-flown language that sounds like a doctoral dissertation on art theory, whereas others tell a simple story in plain language. Do whatever suits you, but keep it brief, on one sheet of paper. At the top, boldly emblazon the words "Artist's Statement" so that people know you're an artist with the confidence to make an artistic statement. You can add a couple of biographical paragraphs at the bottom of your statement that tell about your history in the craft and your life experiences that have contributed to your craft vision.

Woodworking Magazines

Hobbyist woodworking magazines are always on the lookout for interesting projects to run articles on as well as for nice photos of finished projects to show their readers. Getting your work printed in a magazine this way won't cause customers to break your door down, but copies of a magazine with your name in them and pretty pictures of your work are a nice addition to your photo portfolio, and it's fun as well. Got a nice project? Send an editor a photo and a letter. If that magazine isn't interested, don't get discouraged—just try another.

Good Photography

Photography can be a bit of a problem. It's not easy to get a good photo yourself, especially if your work is large, and professional photographers charge a lot of money. However, you must have good photography. The difference between a good photo and a poor one might seem slight to you, but the subjective difference is very large. A good photo leaps out at the viewer and says, "This is a piece of fine work," whereas a poor photo is unappealing. To compound the problem, most craftspeople today pay professionals to get top-notch photos, so that your lower-quality photos pale in comparison. As the saying goes, you gotta pay your dues.

To get into a craft show, you submit an application along with 35-millimeter color slides. The jury for the show looks at all the slides and then decides who's in and who's not. This viewing of the slides makes up the entire decision-making process. The jury is trying to find the most attractive work to make the craft show a success. Providing them with good slides will make the difference between getting into the craft show and not. When the jury sees mediocre slides next to good ones, which do you think they'll pick? Lower-quality crafts with top-quality photos have as good (or better) a chance of being accepted than higher-quality crafts with low-quality photos. That's life in the big city.

What Makes a Good Photo?

A good photo is one that is well exposed—not to bright or too dark—and that does not contain elements that distract the viewer from the beauty of the photographed piece. The photo must be in focus. You want a neutral background so that all you see is the work, not whatever else happens to be behind it. The light needs to be soft, or diffused, so that it doesn't cast sharp shadows. You need to avoid bad reflections on shiny surfaces, such as polished wood or glass, and the whole piece needs to be well lighted so that you can see all its detail.

Hiring a Professional Photographer

A professional photographer will charge you from $300 to $700 to shoot your work. When you look for a pro, don't be afraid to ask hard questions. You're paying a professional rate, so you have a right to expect professional service.

First, ask whether they do commercial work for a living. Commercial work is product photography, which is what you want. Many photographers do weddings, portraiture, and other kinds of work that are different from product photography and require different skills. Such photographers might tell you that they can do your job because they want the work, they have the equipment, and it sounds easy. However, an experienced product photographer will be better equipped and experienced to add or subtract the little nuances that make a big difference in a photo. It's worthwhile to pay a little more for an experienced product photographer.

ORIN'S STORY

Professional woodworker Orin Hutchinson derives a great deal of satisfaction from the work he does in his shop near Santa Cruz, California. He got started in his high school woodshop, where he was best in his class, then went on to renting a series of inexpensive buildings to set up his shop in an old storefront, a Quonset hut, an old chicken coop, and his present barn attic. In these shops he has done a wide variety of dining tables, wall units, stereo cabinets, china hutches, and whatever else customers ask for in wood.

He is very proud of a five-piece set he made for an attorney out of koa wood, including a desk, bookcase, file cabinet, computer desk, and coffee

Can you be present at the shoot? A professional setup includes the capability to take photos that develop instantly, allowing you to see how well the lighting is set up and then to adjust the lighting as necessary before shooting the final film (which takes longer to develop).

Does the photographer have a 4-by-5 camera? A 4-by-5 camera makes negatives that are huge: 4 by 5 inches. This produces a far sharper photograph than do 35-millimeter negatives. If you are paying big bucks for photos, you want the best, so you should get some 4-by-5 shots taken along with the 35-millimeter ones. There are also medium-format cameras, which make a negative that is 2¼ inches square. These are a significant improvement over 35 millimeter, although the 4-by-5 format is still best. There are also 8-by-10 cameras, but this is definite overkill for your purposes.

table. He was given carte blanche to build the pieces as he saw fit, which he did much to the customer's satisfaction.

"What I like about this business is that you don't have an in and out basket that fills up every day. You can build something that you can stand back and appreciate when you are done, and if you did a good job, you know will still be around in 100 years."

Think about what you want before you go for the shoot. What angle do you want to shoot the piece at? Do you want close-up photos of certain areas of the piece? Know what film you want the photographer to use. Get 35-millimeter color slides to use for craft show applications and for making small photos and 4-by-5 color slides to use for making large photos, anything larger than 5 by 7 inches. Below that size, 35-millimeter slides will be plenty sharp; above that size, 35-millimeter shots start to look fuzzy. Get a few black-and-white 4-by-5 slides if you really like your work and want the photos preserved for posterity because color films don't last forever, and black and white does. Have a roll of color print film shot as well if you want to get a lot of cheap photos fast, then have the film developed by any standard photo processing center (one that makes machine prints). Postcards are cheaper than photos per piece, but you'll need to buy more.

Taking Your Own Photos

"But I already have a 35-millimeter camera, so why can't I take my own photos?" you might ask. You can, and if your work is smaller, it's easier to get good photos. Here's how to proceed.

First you need a background. Photo supply stores sell "sweeps," which are large pieces of paper that will cover the entire background to fill the entire photo frame. However, you can use a sheet of butcher paper as well.

Professionals have powerful lighting setups with flash heads that flash light into a reflective umbrella or cloth soft box to diffuse the light before it hits the subject. This diffused light evenly covers the subject and doesn't leave hard shadows. However, camera-mounted flashes can be a problem because if the flash points directly at the subject, the flash is not diffused and sharp shadows result. Also, because the light is emitted so closely to the lens, no soft shadows are left on the surface of the subject. These soft shadows add depth to a photo.

You can deal with these problems by using a bounced camera-mounted flash device, which you attach to your camera. Usually, the device has a tilting feature that lets you point the flash up toward the ceiling. This bounces the light off the ceiling, diffusing it and providing softer shadows on the subject. Mount the camera on a tripod with the flash pointed toward a white ceiling that is of average height (a 15-foot ceiling is too far away). Place the subject in front of the camera on a background and position it the way you want. To get a good exposure with bounced flash, you need to carefully read the flash device's instructions, which will tell you what to do.

You can also take your work outside and shoot it in the bright sun, although the problem here is that you'll get sharp shadows. An outside background of grass and a garden looks better than what's in your garage but is still not as good as having a neutral sweep. Outdoor photography like this is best reserved for items that are intended to be used outdoors, such as deck and lawn furniture, birdhouses, and whirligigs.

Finished Photos in Hand

Once you have your photos, buy a photo album and start putting together your portfolio. Photograph all the work that you can and add to your portfolio over time. You'll find this portfolio to be your best advertising and publicity tool.

A Mini-Course in Crafts-Business Basics

by Barbara Brabec

THIS SECTION OF THE BOOK will familiarize you with important areas of legal and financial concern and enable you to ask the right questions if and when it is necessary to consult with an attorney, accountant, or other business adviser. Although the tax and legal information included here has been carefully researched by the author and is accurate to the best of her knowledge, it is not the business of either the author or publisher to render professional services in the area of business law, taxes, or accounting. Readers should therefore use their own good judgment in determining when the services of a lawyer or other professional would be appropriate to their needs.

Information presented applies specifically to businesses in the United States. However, because many U.S. and Canadian laws are similar, Canadian readers can certainly use the following information as a start-up business plan and guide to questions they need to ask their own local, provincial, or federal authorities.

Contents

1. Starting Right

2. Taxes and Record Keeping

Is Your Activity a "Hobby" or a "Business?"
Self-Employment Taxes
Sales Tax Is Serious Business

3. The Legal Forms of Business

Sole Proprietorship
Partnership
LLC (Limited Liability Company)
Corporation

4. Local and State Laws and Regulations

Business Name Registration
Licenses and Permits
Use of Personal Phone for Business
Zoning Regulations

5. General Business and Financial Information

Making a Simple Business Plan
When You Need an Attorney
Why You Need a Business Checking Account
Accepting Credit Cards

6. Minimizing the Financial Risks of Selling

Selling to Consignment Shops
Selling to Craft Malls
Avoiding Bad Checks

1. Starting Right

In preceding chapters of this book, you learned the techniques of a particular art or craft and realized its potential for profit. You learned what kind of products are likely to sell, how to price them, and how and where you might sell them.

Now that you've seen how much fun a crafts business can be (and how profitable it might be if you were to get serious about selling what you make!) you need to learn about some of the "nitty-gritty stuff" that goes hand-in-hand with even the smallest business based at home. It's easy to start selling what you make and it's satisfying when you earn enough money to make your hobby self-supporting. Many crafters go this far and no further, which is fine. But even a hobby seller must be concerned about taxes and local, state, and federal laws. And if your goal is to build a part- or full-time business at home, you must pay even greater attention to the topics discussed in this section of the book.

Everyone loves to make money . . . but actually starting a business frightens some people because they don't understand what's involved. It's easy to come up with excuses for why we don't do certain things in life; close inspection of those excuses usually boils down to fear of the unknown. We get the shivers when we step out of our comfort zone and try something we've never done before. The simple solution to this problem lies in having the right information at the right time. As someone once said, "Knowledge is the antidote to fear."

The quickest and surest way to dispel fear is to inform yourself about the topics that frighten you. With knowledge comes a sense of power, and that power enables you to move. Whether your goal is merely to earn extra income from your crafts hobby or launch a genuine home-based business, reading the following information will help you get started on the right legal foot, avoid financial pitfalls, and move forward with confidence.

When you're ready to learn more about art or crafts marketing or the operation of a home-based crafts business, a visit to your library or bookstore will turn up many interesting titles. In addition to the special resources listed by this book's author, you will find my list of recommended business books, organizations, periodicals, and other helpful resources in section 10 of this chapter. This information is arranged in a checklist you can use as a plan to get your business up and running.

Before you read my "Mini-Course in Crafts-Business Basics," be assured that I understand where you're coming from because I was once there myself.

For a while I sold my craft work, and this experience led me to write my first book, *Creative Cash*. Now, twenty years later, this crafts-business classic ("my baby") has reached its 6th edition. Few of those who are totally involved in a crafts business today started out with a business in mind. Like me, most began as hobbyists looking for something interesting to do in their spare time, and one thing naturally led to another. I never imagined those many years

Social Security Taxes

When your craft business earnings are more than $400 (net), you must file a Self Employment Tax form (Schedule SE) and pay into your personal Social Security account. This could be quite beneficial for individuals who have some previous work experience but have been out of the workplace for a while. Your re-entry into the business world as a self-employed worker, and the additional contributions to your Social Security account, could result in increased benefits upon retirement.

Because so many senior citizens are starting home-based businesses these days, it should be noted that there is a limit on the amount you can earn before losing Social Security benefits. The good news is that this dollar limit increases every year, and once you are past the age of 70, you can earn any amount of income and still receive full benefits. For more information, contact your nearest Social Security office.

ago when I got serious about my crafts hobby that I was putting my-self on the road to a full-time career as a crafts writer, publisher, author, and speaker. Since I and thousands of others have progressed from hobbyists to professionals, I won't be at all surprised if some-day you, too, have a similar adventure.

2. Taxes and Record Keeping

"Ambition in America is still rewarded . . . with high taxes," the comics quip. Don't you long for the good old days when Uncle Sam lived within his income and without most of yours?

Seriously, taxes are one of the first things you must be concerned about as a new business owner, no matter how small your endeavor. This section offers a brief overview of your tax responsibilities as a sole proprietor.

Is Your Activity a "Hobby" or a "Business?"

Whether you are selling what you make only to get the cost of your supplies back, or actually trying to build a profitable business, you need to understand the legal difference between a profitable hobby and a business, and how each is related to your annual tax return.

The IRS defines a hobby as "an activity engaged in primarily for pleasure, not for profit." Making a profit from a hobby does not automatically place you "in business" in the eyes of the Internal Revenue Service, but the activity will be *presumed* to have been engaged in for profit if it results in a profit in at least three years out of five. Or, to put it another way, a "hobby business" automatically becomes a "real business" in the eyes of the IRS at the point where you can state that you are (1) trying to make a profit, (2) making regular business transactions, and (3) have made a profit three years out of five.

As you know, all income must be reported on your annual tax return. How it's reported, however, has everything to do with the amount of taxes you must pay on this income. If hobby income is under $400, it must be entered on the 1040 tax form, with taxes payable accordingly. If the amount is greater than this, you must file a Schedule C form with your 1040 tax form. This is to your advantage, however, since taxes are due only on your *net profit*. Since you can deduct expenses up to the amount of your hobby income, there may be little or no tax at all on your hobby income.

Self-Employment Taxes

Whereas a hobby cannot show a loss on a Schedule C form, a business can. Business owners must pay not only state and federal income taxes on their profits, but self-employment taxes as well. (See sidebar, "Social Security Taxes," page 213.) Because self-employed people pay Social Security taxes at twice the level of regular, salaried workers, you should strive to lower your annual gross profit figure on the Schedule C form through every legal means possible. One way to do this is through careful record keeping of all expenses related to the operation of your business. To quote IRS publications, expenses are deductible if they are "ordinary, necessary, and somehow connected with the operation and potential profit of your business." In addition to being able to deduct all expenses related to the making and selling of their products, business owners can also depreciate the cost of tools and equipment, deduct the overhead costs of operating a home-based office or studio (called the Home Office Deduction), and hire their spouse or children.

Given the complexity of our tax laws and the fact that they are changing all the time, a detailed discussion of all the tax deductions currently available to small business owners cannot be included in a book of this nature. Learning, however, is as easy as reading a book such as *Small Time Operator* by Bernard Kamoroff (my favorite tax and accounting guide), visiting the IRS Web site, or consulting your regular tax adviser.

You can also get answers to specific tax questions twenty-four hours a day by calling the National Association of Enrolled Agents (NAEA). Enrolled agents (EAs) are licensed by the Treasury Department to represent taxpayers before the IRS. Their rates for doing tax returns are often less than what you would pay for an accountant or CPA. (See my checklist for NAEA's toll-free number you can call to ask for a referral to an EA in your area.)

An important concept to remember is that even the smallest business is entitled to deduct expenses related to its business, and the same tax-saving strategies used by "the big guys" can be used by small business owners. Your business may be small now or still in the dreaming stage, but it could be larger next year and surprisingly profitable a few years from now. Therefore it is in your best interest always to prepare for growth, profit, and taxes by learning all you

Keeping Tax Records

Once you're in business, you must keep accurate records of all income and expenses, but the IRS does not require any special kind of bookkeeping system. Its primary concern is that you use a system that clearly and accurately shows true income and expenses. For the sole proprietor, a simple system consisting of a checkbook, a cash receipts journal, a cash disbursements ledger, and a petty cash fund is quite adequate. Post expenses and income regularly to avoid year-end pile-up and panic.

If you plan to keep manual records, check your local office supply store or catalogs for the *Dome* series of record-keeping books, or use the handy ledger sheets and worksheets included in *Small Time Operator*. (This classic tax and accounting guide by CPA Bernard Kamoroff includes details on how to keep good records and prepare financial reports.) If you have a computer, there are a number of accounting software programs available, such as Intuit Quicken, MYOB (Mind Your Own Business) Accounting, and Intuit Quick-

can about the tax laws and deductions applicable to your business. (See also sidebar, "Keeping Tax Records.")

Sales Tax Is Serious Business

If you live in a state that has a sales tax (all but five states do), and sell products directly to consumers, you are required by law to register with your state's Department of Revenue (Sales Tax division) for a resale tax number. The fee for this in most states ranges from $5 to $25, with some states requiring a bond or deposit of up to $150.

Depending on where you live, this tax number may also be called a Retailer's Occupation Tax Registration Number, resale license, or use tax permit. Also, depending on where you live, the place you must call to obtain this number will have different names. In California, for example, you would contact the State Board of Equalization; in Texas, it's called the State Comptroller's Office.

Books, the latter of which is one of the most popular and best bookkeeping systems for small businesses. The great advantage of computerized accounting is that financial statements can be created at the press of a key after accounting entries have been made.

Regardless which system you use, always get a receipt for everything and file receipts in a monthly envelope. If you don't want to establish a petty cash fund, spindle all of your cash receipts, tally them at month's end, and reimburse your personal outlay of cash with a check written on your business account. On your checkbook stub, document the individual purchases covered by this check.

At year's end, bundle your monthly tax receipt envelopes and file them for future reference, if needed. Since the IRS can audit a return for up to three years after a tax return has been filed, all accounting and tax records should be kept at least this long, but six years is better. Personally, I believe you should keep all your tax returns, journals, and ledgers throughout the life of your business.

Within your state's revenue department, the tax division may have a name such as Sales and Use Tax Division or Department of Taxation and Finance. Generally speaking, if you check your telephone book under "Government," and look for whatever listing comes closest to "Revenue," you can find the right office.

If your state has no sales tax, you will still need a reseller's permit or tax exemption certificate to buy supplies and materials at wholesale prices from manufacturers, wholesalers, or distributors. Note that this tax number is only for supplies and materials used to make your products, not for things purchased at the retail level or for general office supplies.

Once registered with the state, you will begin to collect and remit sales and use tax (monthly, quarterly, or annually, as determined by your state) on all *taxable sales*. This does not mean *all* of your gross income. Different states tax different things. Some states put a sales tax on certain services, but generally you will never have to pay sales tax on income from articles sold to magazines, on teaching or consulting fees, or subscription income (if you happen to publish a newsletter). In addition, sales taxes are not applicable to:

- **items sold on consignment through a charitable organization, shop, or other retail outlet, including craft malls and rent-a-space shops (because the party who sells directly to the consumer is the one who must collect and pay sales tax.)**

- **products you wholesale to others who will be reselling them to consumers. (Be sure to get their tax-exemption ID number for your own files, however, in case you are ever questioned as to why you did not collect taxes on those sales.)**

As you sell throughout the year, your record-keeping system must be set up so you can tell which income is taxable and which is tax-exempt for reporting on your sales tax return.

Collecting Sales Tax at Craft Shows

States are getting very aggressive about collecting sales tax, and agents are showing up everywhere these day, especially at the larger craft fairs, festivals, and small business conferences. As I was writing this chapter, a post on the Internet stated that in New Jersey the sales tax department is routinely contacting show promoters about a month before the show date to get the names and addresses of exhibitors. It is expected that other states will soon be following suit. For this reason, you should always take your resale or tax collection certificate with you to shows.

Although you must always collect sales tax at a show when you sell in a state that has a sales tax, how and when the tax is paid to the state can vary. When selling at shows in other states, you may find that the show promoter has obtained an umbrella sales tax certificate, in which case vendors would be asked to give management a check for sales tax at the end of the show for turning over to a tax agent. Or you may have to obtain a temporary sales tax certificate for a show, as advised by the show promoter. Some sellers who regularly do shows in two or three states say it's easier to get a tax ID number from each state and file an annual return instead of doing taxes on a show-by-show basis. (See sidebar, "Including Tax in the Retail Price," page 220.)

Collecting Sales Tax at a Holiday Boutique

If you're involved in a holiday boutique where several sellers are offering goods to the public, each individual seller will be responsible for collecting and remitting his or her own sales tax. (This means someone has to keep very good records during the sale so each seller receives a record of the sale and the amount of tax on that sale.) A reader who regularly has home boutiques told me that in her community she must also post a sign at her "cash station" stating that sales tax is being collected on all sales, just as craft fair sellers must do in some states. Again, it's important that you get complete details from your own state about its sales tax policies.

Collecting Tax on Internet Sales

Anything you sell that is taxable in your state is also taxable on the Internet. This is simply another method of selling, like craft fairs or mail-order sales. You don't have to break out Internet sales separately; simply include them in your total taxable sales.

3. The Legal Forms of Business

Every business must take one of four legal forms:

Sole Proprietorship
Partnership
LLC (Limited Liability Company)
Corporation

Including Tax in the Retail Price

Is it okay to incorporate the amount of sales tax into the retail price of items being sold directly to consumers? I don't know for sure because each state's sales tax law is different.

Crafters like to use round-figure prices at fairs because this encourages cash sales and eliminates the need for taking coins to make change. Some crafters tell their customers that sales tax has been included in their rounded-off prices, but you should not do this until you check with your state. In some states, this is illegal; in others, you may find that you are required to inform your customers, by means of a sign, that sales tax has been included in your price. Your may also have to print this information on customer receipts as well.

If you make such a statement and collect taxes on cash sales, be sure to report those cash sales as taxable income and remit the tax money to the state accordingly. Failure

As a hobby seller, you automatically become a sole proprietor when you start selling what you make. Although most professional crafters remain sole proprietors throughout the life of their business, some do form craft partnerships or corporations when their business begins to generate serious money, or if it happens to involve other members of their family. You don't need a lawyer to start a sole proprietorship, but it would be folly to enter into a partnership, corporation, or LLC without legal guidance. Here is a brief look at the main advantages and disadvantages of each type of legal business structure.

Sole Proprietorship

No legal formalities are involved in starting or ending a sole proprietorship. You're your own boss here, and the business starts when you say it does and ends automatically when you stop running it. As discussed earlier, income is reported annually on a Schedule C form

to do this would be a violation of the law, and it's easy to get caught these days when sales tax agents are showing up at craft fairs across the country.

Even if rounding off the price and including the tax within that figure turns out to be legal in your state, it will definitely complicate your bookkeeping. For example, if you normally sell an item for $5 or some other round figure, you must have a firm retail price on which to calculate sales tax to begin with. Adding tax to a round figure makes it uneven. Then you must either raise it or lower the price, and if you lower it, what you're really doing is paying the sales tax for your customer out of your profits. This is no way to do business.

I suggest that you set your retail prices based on the pricing formulas given in this book, calculate the sales tax accordingly, and give your customers change if they pay in cash. You will be perceived as a professional when you operate this way, whereas crafters who insist always on "cash only" sales are sending signals to buyers that they don't intend to report this income to tax authorities.

and taxed at the personal level. The sole proprietor is fully liable for all business debts and actions. In the event of a lawsuit, personal assets are not protected.

Partnership

There are two kinds of partnerships: General and Limited

A *General Partnership* is easy to start, with no federal requirements involved. Income is taxed at the personal level and the partnership ends as soon as either partner withdraws from the business. Liability is unlimited. The most financially dangerous thing about a partnership is that the debts incurred by one partner must be assumed by all other partners. Before signing a partnership agreement, make sure the tax obligations of your partner are current.

In a *Limited Partnership,* the business is run by general partners and financed by silent (limited) partners who have no liability beyond an investment of money in the business. This kind of partnership is more complicated to establish, has special tax withholding regulations, and requires the filing of a legal contract with the state.

LLC (Limited Liability Company)

This legal form of business reportedly combines the best attributes of other small business forms while offering a better tax advantage than a limited partnership. It also affords personal liability protection similar to that of a corporation. To date, few craft businesses appear to be using this business form.

Corporation

A corporation is the most complicated and expensive legal form of business and not recommended for any business whose earnings are less than $25,000 a year. If and when your business reaches this point, you should study some books on this topic to fully understand the pros and cons of a corporation. Also consult an accountant or attorney for guidance on the type of corporation you should select—a "C" (general corporation) or an "S" (subchapter S corporation). One book that offers good perspective on this topic is *INC Yourself—How to Profit by Setting Up Your Own Corporation*.

The main disadvantage of incorporation for the small business owner is that profits are taxed twice: first as corporate income and again when they are distributed to the owner-shareholders as dividends. For this reason, many small businesses elect to incorporate as subchapter S corporations, which allows profits to be taxed at owners' regular individual rates. (See sidebar, "The Limited Legal Protection of a Corporation," below.)

The Limited Legal Protection of a Corporation

Business novices often think that by incorporating their business they can protect their personal assets in the event of a lawsuit. This is true if you have employees who do something wrong and cause your business to be sued. As the business owner, however, if you personally do something wrong and are sued as a result, you might in some cases be held legally responsible, and the "corporation door" will offer no legal protection for your personal assets.

Or, as CPA Bernard Kamoroff explains in *Small Time Operator,* "A corporation will not shield you from personal liability that you normally should be responsible for, such as not having car insurance or acting with gross negligence. If you plan to incorporate solely or primarily with the intention of limiting your legal liability, I suggest you find out first exactly how limited the liability really is for your particular venture. Hire a knowledgeable lawyer to give you a written opinion." (See section 7, "Insurance Tips.")

4. Local and State Laws and Regulations

This section will acquaint you with laws and regulations that affect the average art or crafts business based at home. If you've unknowingly broken one of these laws, don't panic. It may not be as bad as you think. It is often possible to get back on the straight and narrow merely by filling out a required form or by paying a small fee of some kind. What's important is that you take steps now to comply with the laws that pertain to your particular business. Often, the fear of being caught when you're breaking a law is often much worse than doing whatever needs to be done to set the matter straight. In the end, it's usually what you don't know that is most likely to cause legal or financial problems, so never hesitate to ask questions about things you don't understand.

Even when you think you know the answers, it can pay to "act dumb." It is said that Napoleon used to attend meetings and pretend to know nothing about a topic, asking many probing questions. By feigning ignorance, he was able to draw valuable information and insight out of everyone around him. This strategy is often used by today's small business owners, too.

Business Name Registration

If you're a sole proprietor doing business under any name other than your own full name, you are required by law to register it on both the local and state level. In this case, you are said to be using an "assumed," "fictitious," or "trade" name. What registration does is enable authorities to connect an assumed name to an individual who can be held responsible for the actions of a business. If you're doing business under your own name, such as Kay Jones, you don't have to register your business name on either the local or state level. If your name is part of a longer name, however (for example, Kay Jones Designs), you should check to see if your county or state requires registration.

Local Registration

To register your name, contact your city or county clerk, who will explain what you need to do to officially register your business on the local level. At the same time, ask if you need any special municipal or county licenses or permits to operate within the law. (See next section, "Licenses and Permits.") This office can also tell you how and where to write to register your name at the state level. If you've been operating under an assumed name for a while and are worried because you didn't register the name earlier, just register it now, as if the business were new.

Registration involves filling out a simple form and paying a small fee, usually around $10 to $25. At the time you register, you will get details about a classified ad you must run in a general-circulation newspaper in your county. This will notify the public at large that you are now operating a business under an assumed name. (If you don't want your neighbors to know what you're doing, simply run the ad in a newspaper somewhere else in the county.) After publication of this ad, you will receive a Fictitious Name Statement that you must send to the County Clerk, who in turn will file it with your registration form to make your business completely legitimate. This name statement or certificate may also be referred to as your DBA ("doing business as") form. In some areas, you cannot open a business checking account if you don't have this form to show your bank.

State Registration

Once you've registered locally, contact your Secretary of State to register your business name with the state. This will prevent its use by a corporate entity. At the same time, find out if you must

Picking a Good Business Name

If you haven't done it already, think up a great name for your new business. You want something that will be memorable—catchy, but not too cute. Many crafters select a simple name that is attached to their first name, such as "Mary's Quilts" or "Tom's Woodcrafts." This is fine for a hobby business, but if your goal is to build a full-time business at home, you may wish to choose a more professional-sounding name that omits your personal name. If a name sounds like a hobby business, you may have difficulty getting wholesale suppliers to take you seriously. A more professional name may also enable you to get higher prices for your products. For example, the above names might be changed to "Quilted Treasures" or "Wooden Wonders."

Don't print business cards or stationery until you find out if someone else is already using the name you've chosen. To find out if the name has already been registered, you

obtain any kind of state license. Generally, home-based craft businesses will not need a license from the state, but there are always exceptions. An artist who built an open-to-the-public art studio on his property reported that the fine in his state for operating this kind of business without a license was $50 a day. In short, it always pays to ask questions to make sure you're operating legally and safely.

Federal Registration

The only way to protect a name on the federal level is with a trademark, discussed in section 8.

Licenses and Permits

A "license" is a certificate granted by a municipal or county agency that gives you permission to engage in a business occupation. A "permit" is similar, except that it is granted by local authorities. Until recently, few craft businesses had to have a license or permit

can perform a trademark search through a search company or hire an attorney who specializes in trademak law to conduct the search for you. And if you are planning to eventually set up a Web site, you might want to do a search to see if that domain name is still available on the Internet. Go to www.networksolutions.com to do this search. Business names have to be registered on the Internet, too, and they can be "parked" for a fee until you're ready to design your Web site.

It's great if your business name and Web site name can be the same, but this is not always possible. A crafter told me recently she had to come up with 25 names before she found a domain name that hadn't already been taken. (Web entrepreneurs are grabbing every good name they can find. Imagine my surprise when I did a search and found that two different individuals had set up Web sites using the titles of my two best-known books, *Creative Cash* and *Homemade Money*.)

of any kind, but a growing number of communities now have new laws on their books that require home-based business owners to obtain a "home occupation permit." Annual fees for such permits may range from $15 to $200 a year. For details about the law in your particular community or county, call your city or country clerk (depending on whether you live within or outside city limits).

Use of Personal Phone for Business

Although every business writer stresses the importance of having a business telephone number, craftspeople generally ignore this advice and do business on their home telephone. While it's okay to use a home phone to make outgoing business calls, you cannot advertise a home telephone number as your business phone number without being in violation of local telephone regulations. That means you cannot legally put your home telephone number on a business card or business stationery or advertise it on your Web site.

That said, let me also state that most craftspeople totally ignore this law and do it anyway. (I don't know what the penalty for breaking this law is in your state; you'll have to call your telephone company for that information and decide if this is something you want to do.) Some phone companies might give you a slap on the wrist and tell you to stop, while others might start charging you business line telephone rates if they discover you are advertising your personal phone number.

The primary reason to have a separate phone line for your business is that it enables you to freely advertise your telephone number to solicit new business and invite credit card sales, custom order inquiries, and the like. Further, you can deduct 100 percent of the costs of a business telephone line on your Schedule C tax form, while deductions for the business use of a home phone are severely limited. (Discuss this with your accountant.)

If you plan to connect to the Internet or install a fax machine, you will definitely need a second line to handle the load, but most crafters simply add an additional personal line instead of a business line. Once on the Internet, you may have even less need for a business phone than before since you can simply invite contact from buyers by advertising your e-mail address. (Always include your e-mail and Internet address on your business cards and stationery.)

If your primary selling methods are going to be consignment shops, craft fairs, or craft malls, a business phone number would be necessary only if you are inviting orders by phone. If you present a holiday boutique or open house once or twice a year, there should be no problem with putting your home phone number on promotional fliers because you are, in fact, inviting people to your home and not your business (similar to running a classified ad for a garage sale).

If and when you decide a separate line for your business is necessary, you may find it is not as costly as you think. Telephone companies today are very aware of the number of people who are working at home, and they have come up with a variety of affordable packages and second-line options, any one of which might be perfect for your craft business needs. Give your telephone company a call and see what's available.

Zoning Regulations

Before you start any kind of home-based business, check your home's zoning regulations. You can find a copy at your library or at city hall. Find out what zone you're in and then read the information under "home occupations." Be sure to read the fine print and note the penalty for violating a zoning ordinance. In most cases, someone who is caught violating zoning laws will be asked to cease and desist and a penalty is incurred only if this order is ignored. In other cases, however, willful violation could incur a hefty fine.

Zoning laws differ from one community to another, with some of them being terribly outdated (actually written back in horse-and-buggy days). In some communities, zoning officials simply "look the other way" where zoning violations are concerned because it's easier to do this than change the law. In other places, however, zoning regulations have recently been revised in light of the growing number of individuals working at home, and these changes have not always been to the benefit of home-based workers or self-employed individuals. Often there are restrictions as to (1) the amount of space in one's home a business may occupy (impossible to enforce, in my opinion), (2) the number of people (customers, students) who can come to your home each day, (3) the use of non-family employees, and so on. If you find you cannot advertise your home as a place of business, this problem can be easily solved by renting a P.O. box or using a commercial mailbox service as your business address.

Although I'm not suggesting that you violate your zoning law, I will tell you that many individuals who have found zoning to be a problem do ignore this law, particularly when they have a quiet business that is unlikely to create problems in their community.

Zoning officials don't go around checking for people who are violating the law; rather, they tend to act on complaints they have received about a certain activity that is creating problems for others. Thus, the best way to avoid zoning problems is to keep a low profile by not broadcasting your home-based business to neighbors. More important, never annoy them with activities that emit fumes or odors, create parking problems, or make noise of any kind.

While neighbors may grudgingly put up with a noisy hobby activity (such as sawing in the garage), they are not likely to tolerate the same noise or disturbance if they know it's related to a home-based business. Likewise, they won't mind if you have a garage sale every year, but if people are constantly coming to your home to buy from your home shop, open house, home parties, or holiday boutiques every year, you could be asking for trouble if the zoning laws don't favor this kind of activity.

5. General Business and Financial Information

This section offers introductory guidelines on essential business basics for beginners. Once your business is up and running, however, you need to read other craft-business books to get detailed information on the following topics and many others related to the successful growth and development of a home-based art or crafts business.

Making a Simple Business Plan

As baseball star Yogi Berra once said, "If you don't know where you are going, you might not get there." That's why you need a plan.

Like a road map, a business plan helps you get from here to there. It doesn't have to be fancy, but it does have to be in written form. A good business plan will save you time and money while helping you stay focused and on track to meet your goals. The kind of business plan a craftsperson makes will naturally be less complicated than the business plan of a major manufacturing company, but the elements are basically the same and should include:

- *History*—how and why you started your business
- *Business description*—what you do, what products you make, why they are special
- *Management information*—your business background or experience and the legal form your business will take
- *Manufacturing and production*—how and where products will be produced and who will make them; how and where supplies and materials will be obtained, and their estimated costs; labor costs (yours or other helpers); and overhead costs involved in the making of products
- *Financial plan*—estimated sales and expense figures for one year
- *Market research findings*—a description of your market (fairs, shops, mail order, Internet, etc.), your customers, and your competition
- *Marketing plan*—how you are going to sell your products and the anticipated cost of your marketing (commissions, advertising, craft fair displays, etc.)

If this all seems a bit much for a small crafts business, start managing your time by using a daily calendar/planner and start a notebook you can fill with your creative and marketing ideas, plans, and business goals. In it, write a simple mission statement that answers the following questions:

- What is my primary mission or goal in starting a business?
- What is my financial goal for this year?
- What am I going to do to get the sales I need this year to meet my financial goal?

The most important thing is that you start putting your dreams, goals, and business plans on paper so you can review them regularly.

It's always easier to see where you're going if you know where you've been.

When You Need an Attorney

Many business beginners think they have to hire a lawyer the minute they start a business, but that would be a terrible waste of money if you're just starting a simple art or crafts business at home, operating as a sole proprietor. Sure, a lawyer will be delighted to hold your hand and give you the same advice I'm giving you here (while charging you $150 an hour or more for his or her time). With this book in hand, you can easily take care of all the "legal details" of small business start-up. The day may come, however, when you do need legal counsel, such as when you:

Form a Partnership or Corporation

As stated earlier, an attorney's guidance is necessary in the formation of a partnership. Although many people have incorporated without a lawyer using a good how-to book on the topic, I wouldn't recommend doing this because there are so many details involved here, not to mention different types of corporate entities.

Defend an Infringement of a Copyright or Trademark

You don't need an attorney to get a simple copyright, but if someone infringes on one of your copyrights, you will probably need legal help to stop the infringer from profiting from your creativity. You can file your own trademark application (if you are exceedingly careful about following instructions), but it would be difficult to protect your trademark without legal help if someone tries to steal it. In both cases, you would need an attorney who specializes in copyright, patent, and trademark law. (If you ever need a good attorney who understands the plight of artists and crafters, contact me by e-mail at barbara@crafter.com and I'll refer you to

Get a Safety Deposit Box

The longer you are in business, the more important it will be to safeguard your most valuable business records. When you work at home, there is always the possibility of fire or damage from some natural disaster, be it tornado, earthquake, hurricane, or flood. You will worry less if you keep your most valuable business papers, records, computer disks, and so forth off-premises, along with other items that would be difficult or impossible to replace. Some particulars I have always kept in my business safety deposit box include master software disks and computer back-up tapes; original copies of my designs and patterns, business contracts, copyrights, insurance policies, and a photographic record of all items insured on our homeowner's policy. Remember: Insurance is worthless if you cannot prove what you owned in the first place.

the attorney who has been helpful to me in protecting my common-law trademark to *Homemade Money*, my home-business classic. The 6th edition of this book includes the details of my trademark infringement story.)

Negotiate a Contract

Many craft hobbyists of my acquaintance have gone on to write books and sell their original designs to manufacturers, suddenly finding themselves with a contract in hand that contains a lot of confusing legal jargon. When hiring an attorney to check any kind of contract, make sure he or she has experience in the particular field involved. For example, a lawyer specializing in real estate isn't going to know a thing about the inner workings of a book publishing company and how the omission or inclusion of a particular clause or phrase might impact the author's royalties or make it difficult to get publishing rights back when the book goes out of print. Although I have no experience in the licensing industry, I presume the same thing holds true here. What I do know for sure is that the problem with most contracts is not so much what's *in* them, as what *isn't.* Thus you need to be sure the attorney you hire for specialized contract work has done this kind of work for other clients.

Hire Independent Contractors

If you ever grow your business to the point where you need to hire workers and are wondering whether you have to hire employees or can use independent contractors instead, I suggest you to seek counsel from an attorney who specializes in labor law. This topic is very complex and beyond the scope of this beginner's guide, but I do want you to know that the IRS has been on a campaign for the past several years to abolish independent contractors altogether. Many small businesses have suffered great financial loss in back taxes and penalties because they followed the advice of an accountant or regular attorney who didn't fully understand the technicalities of this matter.

If and when you do need a lawyer for general business purposes, ask friends for a reference, and check with your bank, too, since it will probably know most of the attorneys with private practices in your area. Note that membership in some small business organizations will also give you access to affordable prepaid legal services. If you ever need serious legal help but have no funds to pay for it, contact the Volunteer Lawyers for the Arts (see resources in section 10).

Why You Need a Business Checking Account

Many business beginners use their personal checking account to conduct the transactions of their business, *but you must not do this* because the IRS does not allow co-mingling of business and personal income. If you are operating as a business, reporting income on a Schedule C form and taking deductions accordingly, the lack of a separate checking account for your business would surely result in an IRS ruling that your endeavor was a hobby and not a business. That, in turn, would cost you all the deductions previously taken on earlier tax returns and you'd end up with a very large tax bill. Don't you agree that the cost of a separate checking account is a small price to pay to protect all your tax deductions?

You do not necessarily need one of the more expensive business checking accounts; just a *separate account* through which you run all business income and expenditures. Your business name does not have to be on these checks so long as only your name (not your spouse's) is listed as account holder. You can save money on your checking account by first calling several banks and savings and loan institutions and comparing the charges they set for imprinted checks, deposits, checks written, bounced checks, and other services. Before you open your account, be sure to ask if the bank can set you up to take credit cards (merchant account) at some point in the future.

Accepting Credit Cards

Most of us today take credit cards for granted and expect to be able to use them for most everything we buy. It's nice to be able to offer credit card services to your craft fair customers, but it is costly and thus not recommended for beginning craft sellers. If you get into selling at craft fairs on a regular basis, however, at some point you may find you are losing sales because you don't have "merchant status" (the ability to accept credit cards as payment).

Some craftspeople have reported a considerable jump in sales once they started taking credit cards. That's because some people who buy with plastic may buy two or three items instead of one, or are willing to pay a higher price for something if they can charge it. Thus, the higher your prices, the more likely you are to lose sales if you can't accept credit cards. As one jewelry maker told me, "I always seem to get the customers who have run out of cash and left their checkbook at home. But even when they have a check, I feel uncomfortable taking a check for $100 or more."

A list follows of the various routes you can travel to get merchant status. You will have to do considerable research to find out which method is best for you. All will be costly, and you must have sufficient sales, or the expectation of increased sales, to consider taking credit cards in the first place. Understand, too, that taking credit cards in person (called face-to-face transactions where you have the card in front of you) is different from accepting credit cards by phone, by mail, or through a Web site (called non–face-to-face transactions). Each method of selling is treated differently by bankcard providers.

Merchant Status from Your Bank

When you're ready to accept credit cards, start with the bank where you have your business checking account. Where you bank, and where you live, has everything to do with whether you can get merchant status from your bank or not. Home-business owners in small towns often have less trouble than do those in large cities. One crafter told me Bank of America gave her merchant status with no problem, but some banks simply refuse to deal with anyone who doesn't operate out of a storefront. Most banks now insist that credit card sales be transmitted electronically, but a few still offer manual printers and allow merchants to send in their sales slips by mail. You will be given details about this at the time you apply for merchant status. All banks will require proof that you have a going business and will want to see your financial statements.

Merchant Status through a Crafts Organization

If you are refused by your bank because your business is home based or just too new, getting bankcard services through a crafts or home-business organization is the next best way to go. Because such organizations have a large membership, they have some negotiating power with the credit card companies and often get special deals for their members. As a member of such an organization, the chances are about 95 percent that you will automatically be accepted into an its bankcard program, even if you are a brand new business owner.

One organization I can recommend to beginning sellers is the National Craft Association. Managing Director Barbara Arena tells me that 60 percent of all new NCA members now take the MasterCard/VISA services offered by her organization. "Crafters who are unsure about whether they want to take credit cards over a long period of time have the option of renting equipment," says Barbara. "This enables them to get out of the program with a month's notice. NCA members can operate on a software basis through their personal computer (taking their laptop computer to shows and calling in sales on their cell phone), or use a swipe machine. Under NCA's program, crafters can also accept credit card sales on their Internet site."

For more information from NCA and other organizations offering merchant services, see "Craft and Home-Business Organizations" on page 278.

Merchant Status from Credit Card Companies

If you've been in business for a while, you may find you can get merchant status directly from American Express or Novus Services, Inc., the umbrella company that handles the Discover, Bravo, and Private Issue credit cards. American Express says that in some cases it can grant merchant status immediately upon receipt of some key information given on the phone. As for Novus, many crafters have told me how easy it was to get merchant status from this company. Novus says it only needs your Social Security number and information to check your credit rating. If Novus accepts you, it can also get you set up to take VISA and MasterCard as well if you meet the special acceptance qualifications of these two credit card companies. (Usually, they require you to be in business for at least two years.)

Merchant Status from an Independent
Service Organization Provider (ISO)

ISOs act as agents for banks that authorize credit cards, promoting their services by direct mail, through magazine advertising, telemarketing, and on the Internet. Most of these bankcard providers are operating under a network marketing program (one agent representing one agent representing another, and so on). They are everywhere on the Internet, sending unsolicited e-mail messages to Web site owners. In addition to offering the merchant account service itself, many are also trying to get other Web site owners to promote the same service in exchange for some kind of referral fee. I do not recommend that you get merchant status through an ISO because I've heard too many horror stories about them. If you want to explore this option on the Internet, however, use your browser's search button and type "credit cards + merchant" to get a list of such sellers.

In general, ISOs may offer a low discount rate but will sock it to you with inflated equipment costs, a high application fee, and extra fees for installation, programming, and site inspection. You will also have to sign an unbreakable three- or four-year lease for the electronic equipment.

As you can see, you must really do your homework where bankcard services are concerned. In checking out the services offered by any of the providers noted here, ask plenty of questions. Make up a chart that lets you compare what each one charges for application and service fees, monthly charges, equipment costs, software, discount rates, and transaction fees.

Transaction fees can range from 20 to 80 cents per ticket, with discount rates running anywhere from 1.67 percent to 5 percent. Higher rates are usually attached to non–face-to-face credit card transactions, paper transaction systems, or a low volume of sales. Any rate higher than 5 percent should be a danger signal since you could be dealing with an unscrupulous seller or some kind of illegal third-party processing program.

I'm told that a good credit card processor today may cost around $800, yet some card service providers are charging two or three times that amount in their leasing arrangements. I once got a quote from a major ISO and found it would have cost me $40 a month to lease the terminal—$1,920 over a period of four years—or I could buy it for just $1,000. In checking with my bank, I learned I could get the same equipment and the software to run it for just $350!

In summary, if you're a nervous beginner, the safest way to break into taking credit cards is to work with a bank or organization that offers equipment on a month-by-month rental arrangement. Once you've had some experience in taking credit card payments, you can review your situation and decide whether you want to move into a leasing arrangement or buy equipment outright.

6. Minimizing the Financial Risks of Selling

This book contains a good chapter on how and where to sell your crafts, but I thought it would be helpful for you to have added perspective on the business management end of selling through various outlets, and some things you can do to protect yourself from financial loss and legal hassles.

First you must accept the fact that all businesses occasionally suffer financial losses of one kind or another. That's simply the nature of business. Selling automatically carries a certain degree of risk in that we can never be absolutely sure that we're going to be paid for anything until we actually have payment in hand. Checks may bounce, wholesale buyers may refuse to pay their invoices, and consignment shops can close unexpectedly without returning merchandise to crafters. In the past few years, a surprising number

State Consignment Laws

Technically, consigned goods remain the property of the seller until they are sold. When a shop goes out of business, however, consigned merchandise may be seized by creditors in spite of what your consignment agreement may state. You may have some legal protection here, however, if you live in a state that has a consignment law designed to protect artists and craftspeople in such instances. I believe such laws exist in the states of CA, CO, CT, IL, IA, KY, MA, NH, NM, NY, OR, TX, WA, and WI. Call your Secretary of State to confirm this or, if your state isn't listed here, ask whether this law is now on the books. Be sure to get full details about the kind of protection afforded by this law because some states have different definitions for what constitutes "art" or "crafts."

of craft mall owners have stolen out of town in the middle of the night, taking with them all the money due their vendors, and sometimes the vendors' merchandise as well. (This topic is beyond the scope of this book, but if you'd like more information on it, see my *Creative Cash* book and back issues of my *Craftsbiz Chat* newsletter on the Internet at www.crafter.com/brabec).

Now I don't want you to feel uneasy about selling or suspicious of every buyer who comes your way, because that would take all the fun out of selling. But I *do* want you to know that bad things sometimes happen to good craftspeople who have not done their homework (by reading this book, you are doing *your* homework). If you will follow the cautionary guidelines that follow, you can avoid some common selling pitfalls and minimize your financial risk to the point where it will be negligible.

Selling to Consignment Shops

Never consign more merchandise to one shop than you can afford to lose, and do not send new items to a shop until you see that payments are being made regularly according to your written consignment agreement. It should cover the topics of:

- insurance (see "Insurance Tips," section 7).
- pricing (make sure the shop cannot raise or lower your retail price without your permission).
- sales commission (40 percent is standard; don't work with shop owners who ask for more than this. It makes more sense to wholesale products at 50 percent and get payment in 30 days).
- payment dates.
- display of merchandise.
- return of unsold merchandise (some shops have a clause stating that if unsold merchandise is not claimed within 30 to 60 days after a notice has been sent, the shop can dispose of it any way it wishes).

Above all, make sure your agreement includes the name and phone number of the shop's owner (not just the manager). If a shop fails and you decide to take legal action, you want to be sure your lawyer can track down the owner. (See sidebar, "State Consignment Laws," page 240.)

Selling to Craft Malls

Shortly after the craft mall concept was introduced to the crafts community in 1988 by Rufus Coomer, entrepreneurs who understood the profit potential of such a business began to open malls all over the country. But there were no guidebooks and everyone was flying by the seat of his or her pants, making up operating rules along the way. Many mall owners, inexperienced in retailing, have since gone out of business, often leaving crafters holding the bag. The risks of selling through such well-known chain stores as Coomers or American Craft Malls are minimal, and many independently owned malls have also established excellent reputations in the industry. What you need to be especially concerned about here are new malls opened by individuals who have no track record in this industry.

I'm not telling you *not* to set up a booth in a new mall in your area—it might prove to be a terrific outlet for you—but I am cautioning you to keep a sharp eye on the mall and how it's being operated. Warning signs of a mall in trouble include:

- **less than 75 percent occupancy**
- **little or no ongoing advertising**
- **not many shoppers**
- **crafters pulling out (usually a sign of too few sales)**
- **poor accounting of sales**
- **late payments**

If a mall is in trouble, it stands to reason that the logical time for it to close is right after the biggest selling season of the year, namely Christmas. Interestingly, this is when most of the shady mall owners have stolen out of town with crafters' Christmas sales in their pockets. As stated in my *Creative Cash* book:

> If it's nearing Christmas time, and you're getting uncomfortable vibes about the financial condition of a mall you're in, it might be smart to remove the bulk of your merchandise—especially expensive items—just before it closes for the holidays. You can always restock after the first of the year if everything looks rosy.

Avoiding Bad Checks

At a crafts fair or other event where you're selling directly to the public, if the buyer doesn't have cash and you don't accept credit cards, your only option is to accept a check. Few crafters have bad check problems for sales held in the home (holiday boutique, open house, party plan, and such), but bad checks at craft fairs are always possible. Here are several things you can do to avoid accepting a bad check:

- Always ask to see a driver's license and look carefully at the picture on it. Write the license number on the check.

- If the sale is a for a large amount, you can ask to see a credit card for added identification, but writing down the number will do no good because you cannot legally cover a bad check with a customer's credit card. (The customer has a legal right to refuse to let you copy the number as well.)

- Look closely at the check itself. Is there a name and address printed on it? If not, ask the customer to write in this information by hand, along with his or her phone number.

- Look at the sides of the check. If at least one side is not perforated, it could be a phony check.

- Look at the check number in the upper right-hand corner. Most banks who issue personalized checks begin the numbering system with 101 when a customer reorders new checks. The Small Business Administration says to be more cautious with low sequence numbers because there seems to be a higher number of these checks that are returned.

- Check the routing number in the lower left-hand corner and note the ink. If it looks shiny, wet your finger and see if the ink rubs off. That's a sure sign of a phony check because good checks are printed with magnetic ink that does not reflect light.

Collecting on a Bad Check

No matter how careful you are, sooner or later, you will get stuck with a bad check. It may bounce for three reasons:

> nonsufficient funds (NSF)
> account closed
> no account (evidence of fraud)

I've accepted tens of thousands of checks from mail-order buyers through the years and have rarely had a bad check I couldn't collect with a simple phone call asking the party to honor his or her obligation to me. People often move and close out accounts before all checks have cleared, or they add or subtract wrong, causing their account to be overdrawn. Typically, they are embarrassed to have caused a problem like this.

When the problem is more difficult than this, your bank can help. Check to learn its policy regarding bounced checks. Some automatically put checks through a second time. If a check bounces at this point, you may ask the bank to collect the check for you. The check needs to be substantial, however, since the bank fee may be $15 or more if they are successful in collecting the money.

If you have accepted a check for a substantial amount of money and believe there is evidence of fraud, you may wish to do one of the following:

- notify your district attorney's office
- contact your sheriff or police department (since it is a crime to write a bad check)
- try to collect through small claims court

For more detailed information on all of these topics, see *The Crafts Business Answer Book*.

7. Insurance Tips

As soon as you start even the smallest business at home, you need to give special attention to insurance. This section offers an introductory overview of insurance concerns of primary interest to crafts-business owners.

Homeowner's or Renter's Insurance

Anything in the home being used to generate income is considered to be business-related and thus exempt from coverage on a personal policy. Thus your homeowner's or renter's insurance policy will not cover business equipment, office furniture, supplies, or inventory of finished goods unless you obtain a special rider. Such riders, called a "Business Pursuits Endorsement" by some companies, are inexpensive and offer considerable protection. Your insurance agent will be happy to give you details.

As your business grows and you have an ever-larger inventory of supplies, materials, tools, and finished merchandise, you may find it necessary to buy a special in-home business policy that offers broader protection. Such policies may be purchased directly from insurance companies or through craft and home-business organizations that offer special insurance programs to their members.

Liability Insurance

There are two kinds of liability insurance. *Product* liability insurance protects you against lawsuits by consumers who have been injured while using one of your products. *Personal* liability insurance protects you against claims made by individuals who have suffered bodily injury while on your premises (either your home or the place where you are doing business, such as in your booth at a crafts fair).

Your homeowner's or renter's insurance policy will include some personal liability protection, but if someone were to suffer bodily injury while on your premises for *business* reasons, that coverage might not apply. Your need for personal liability insurance will be greater if you plan to regularly present home parties, holiday boutiques, or open house sales in your home where many people might be coming and going throughout the year. If you sell at craft fairs, you would also be liable for damages if someone were to fall and be injured in your booth or if something in your booth falls and injures another person. For this reason, some craft fair promoters now require all vendors to have personal liability insurance.

As for product liability insurance, whether you need it or not depends largely on the type of products you make for sale, how careful you are to make sure those products are safe, and how and where you sell them. Examples of some crafts that have caused injury to consumers and resulted in court claims in the past are stuffed toys with wire or pins that children have swallowed; items made of yarn or fiber that burned rapidly; handmade furniture that collapsed when someone put an ordinary amount of weight on them; jewelry with sharp points or other features that cut the wearer, and so on. Clearly, the best way to avoid injury to consumers is to make certain your products have no health hazards and are safe to use. (See discussion of consumer safety laws in section 8.)

Few artists and craftspeople who sell on a part-time basis feel they can afford product liability insurance, but many full-time craft professionals, particularly those who sell their work wholesale, find it a necessary expense. In fact, many wholesale buyers refuse to buy from suppliers that do not carry product liability insurance.

I believe the least expensive way to obtain both personal and product liability insurance is with one of the comprehensive in-home or craft business policies offered by a craft or home-business organization. Such policies generally offer a million dollars of both personal and product liability coverage. (See "Things to Do" Checklist on page 273 and Resources for some organizations you can contact for more information. Also check with your insurance agent about the benefits of an umbrella policy for extra liability insurance.)

Insurance on Crafts Merchandise

As a seller of art or crafts merchandise, you are responsible for insuring your own products against loss. If you plan to sell at craft fairs, in craft malls, rent-a-space shops, or consignment shops, you may want to buy an insurance policy that protects your merchandise both at home or away. Note that while craft shops and malls generally have fire insurance covering the building and its fixtures, this coverage cannot be extended to merchandise offered for sale because it is not the property of the shop owner. (Exception: Shops and malls in shopping centers are mandated by law to buy fire insurance on their contents whether they own the merchandise or not.)

This kind of insurance is usually part of the home-business/ crafts-business insurance policies mentioned earlier.

Auto Insurance

Be sure to talk to the agent who handles your car insurance and explain that you may occasionally use your car for business purposes. Normally, a policy issued for a car that's used only for pleasure or driving to and from work may not provide complete coverage for an accident that occurs during business use of the car, particularly if the insured is to blame for the accident. For example, if you were delivering a load of crafts to a shop or on your way to a crafts fair and had an accident, would your business destination and the "commercial merchandise" in your car negate your coverage in any

Insuring Your Art or Crafts Collection

The replacement cost insurance you may have on your personal household possessions does not extend to "fine art," which includes such things as paintings, antiques, pictures, tapestries, statuary, and other articles that cannot be replaced with new articles. If you have a large collection of art, crafts, memorabilia, or collector's items, and its value is more than $1,500, you may wish to have your collection appraised so it can be protected with a separate all-risk endorsement to your homeowner's policy called a "fine arts floater."

way? Where insurance is concerned, the more questions you ask, the better you'll feel about the policies you have.

8. Important Regulations Affecting Artists and Craftspeople

Government agencies have a number of regulations that artists and craftspeople must know about. Generally, they relate to consumer safety, the labeling of certain products and trade practices. Following are regulations of primary interest to readers of books in the For Fun & Profit series. If you find a law or regulation related to your particular art or craft interest, be sure to request additional information from the government agency named there.

Consumer Safety Laws

All product sellers must pay attention to the Consumer Product Safety Act, which protects the public against unreasonable risks of injury associated with consumer products. The Consumer Product Safety Commission (CPSC) is particularly active in the area of toys and consumer goods designed for children. All sellers of handmade products must be doubly careful about the materials they use for children's products since consumer lawsuits are common where products for children are concerned. To avoid this problem, simply comply with the consumer safety laws applicable to your specific art or craft.

Toy Safety Concerns

To meet CPSC's guidelines for safety, make sure any toys you make for sale are:

- too large to be swallowed
- not apt to break easily or leave jagged edges

- free of sharp edges or points
- not put together with easily exposed pins, wires, or nails
- nontoxic, nonflammable, and nonpoisonous

The Use of Paints, Varnishes, and Other Finishes

Since all paint sold for household use must meet the Consumer Product Safety Act's requirement for minimum amounts of lead, these paints are deemed to be safe for use on products made for children, such as toys and furniture. Always check, however, to make sure the label bears a nontoxic notation. Specialty paints must carry a warning on the label about lead count, but "artist's paints" are curiously exempt from CPS's lead-in-paint ban and are not required to bear a warning label of any kind. Thus you should *never* use such paints on products intended for use by children unless the label specifically states they are *nontoxic* (lead-free). Acrylics and other water-based paints, of course, are nontoxic and completely safe for use on toys and other products made for children. If you plan to use a finishing coat, make sure it is nontoxic as well.

Fabric Flammability Concerns

The Flammable Fabrics Act is applicable only to those who sell products made of fabric, particularly products for children. It prohibits the movement in interstate commerce of articles of wearing apparel and fabrics that are so highly flammable as to be dangerous when worn by individuals, and for other purposes. Most fabrics comply with the above act, but if you plan to sell children's clothes or toys, you may wish to take an extra step to be doubly sure the fabric you are using is safe. This is particularly important if you plan to wholesale your products. What you should do is ask your fabric supplier for a *guarantee of compliance with the Flammability Act*. This guarantee is generally passed along to the buyer by a statement on the invoice that reads "continuing guaranty under the Flammable Fabrics Act." If you do not find such a statement on your invoice, you should ask the fabric manufacturer, wholesaler, or distributor to furnish you with their "statement of compliance" with the flammability standards. The CPSC can also tell you if a particular manufacturer has filed a continuing guarantee under The Flammable Fabrics Act.

Labels Required by Law

The following information applies only to crafters who use textiles, fabrics, fibers, or yarn products to make wearing apparel, decorative accessories, household furnishings, soft toys, or any product made of wool.

Different governmental agencies require the attachment of certain tags or labels to products sold in the consumer marketplace, whether manufactured in quantity or handmade for limited sale. You don't have to be too concerned about these laws if you sell only at local fairs, church bazaars, and home boutiques. As soon as you get out into the general consumer marketplace, however—doing large craft fairs, selling through consignment shops, craft malls, or wholesaling to shops—it would be wise to comply with all the federal labeling laws. Actually, these laws are quite easy to comply with because the required labels are readily available at inexpensive prices, and you can even make your own if you wish. Here is what the federal government wants you to tell your buyers in a tag or label:

- **What's in a product, and who has made it.** The Textile Fiber Products Identification Act (monitored both by the Bureau of Consumer Protection and the Federal Trade Commission) requires that a special label or hangtag be attached to all textile wearing apparel and household furnishings, with the exception of wall hangings. "Textiles" include products made of any fiber, yarn, or fabric, including garments and decorative accessories, quilts, pillows, place mats, stuffed toys, rugs, etc. The tag or label must include

(1) the name of the manufacturer and (2) the generic names and percentages of all fibers in the product in amounts of 5 percent or more, listed in order of predominance by weight.

■ How to take care of products. Care Labeling Laws are part of the Textile Fiber Products Identification Act, details about which are available from the FTC. If you make wearing apparel or household furnishings of any kind using textiles, suede, or leather, you must attach a permanent label that explains how to take care of the item. This label must indicate whether the item is to be dry-cleaned or washed. If it is washable, you must indicate whether in hot or cold water, whether bleach may or may not be used, and the temperature at which it may be ironed. (See sample labels in sidebar)

■ Details about products made of wool. If a product contains wool, the FTC requires additional identification under a separate law known as the Wool Products Labeling Act of 1939. FTC rules require that the labels of all wool or textile products clearly indicate when imported ingredients are used. Thus, the label for a skirt knitted in the U.S. from wool yarn imported from England would read, "Made in the USA from imported products" or similar wordage. If the wool yarn was spun in the U.S., a product made from that yarn would simply need a tag or label stating it was "Made in the USA" or "Crafted in USA" or some similarly clear terminology.

The Bedding and Upholstered Furniture Law

This is a peculiar state labeling law that affects sellers of items that have a concealed filling. It requires the purchase of a license, and products must have a tag that bears the manufacturer's registry number.

Bedding laws have long been a thorn in the side of crafters because they make no distinction between the large manufacturing company that makes mattresses and pillows, and the individual crafts producer who sells only handmade items. "Concealed filling" items include not just bedding and upholstery, but handmade pillows and quilts. In some states, dolls, teddy bears, and stuffed soft sculpture items are also required to have a tag.

Fortunately, only twenty-nine states now have this law on the books, and even if your state is one of them, the law may be arbitrarily enforced. (One exception is the state of Pennsylvania, which is reportedly sending officials to craft shows to inspect merchandise to see if it is properly labeled.) The only penalty that appears to be connected with a violation of this law in any state is removal of merchandise from store shelves or craft fair exhibits. That being the case, many crafters choose to ignore this law until they are challenged. If you learn you must comply with this law, you will be required to obtain a state license that will cost between $25 and $100, and you will have to order special "bedding stamps" that can be attached to your products. For more information on this complex topic, see *The Crafts Business Answer Book*.

FTC Rule for Mail-Order Sellers

Even the smallest home-based business needs to be familiar with Federal Trade Commission (FTC) rules and regulations. A variety of free booklets are available to business owners on topics related to advertising, mail-order marketing, and product labeling (as discussed earlier). In particular, crafters who sell by mail need to pay attention to the FTC's Thirty-Day Mail-Order Rule, which states that one must ship customer orders within thirty days of receiving payment for the order. This rule is strictly enforced, with severe financial penalties for each violation.

Unless you specifically state in your advertising literature how long delivery will take, customers will expect to receive the product within thirty days after you get their order. If you cannot meet this shipping date, you must notify the customer accordingly, enclosing a postage-paid reply card or envelope, and giving them the option to cancel the order if they wish. Now you know why so many catalog sellers state, "Allow six weeks for delivery." This lets them off the hook in case there are unforeseen delays in getting the order delivered.

9. Protecting Your Intellectual Property

"Intellectual property," says Attorney Stephen Elias in his book, *Patent, Copyright & Trademark,* "is a product of the human intellect that has commercial value."

This section offers a brief overview of how to protect your intellectual property through patents and trademarks, with a longer discussion of copyright law, which is of the greatest concern to individuals who sell what they make. Since it is easy to get patents, trademarks, and copyrights mixed up, let me briefly define them for you:

- A *patent* is a grant issued by the government that gives an inventor the right to exclude all others from making, using, or selling an invention within the United States and its territories and possessions.

- A *trademark* is used by a manufacturer or merchant to identify his or her goods and distinguish them from those manufactured or sold by others.

- A *copyright* protects the rights of creators of intellectual property in five main categories (described in this section).

Perspective on Patents

A patent may be granted to anyone who invents or discovers a new and useful process, machine, manufacture or composition of matter, or any new and useful improvement thereof. Any new, original, and ornamental design for an article of manufacture can also be patented. The problem with patents is that they can cost as much as

$5,000 or more to obtain, and once you've got one, they still require periodic maintenance through the U.S. Patent and Trademark Office. To contact this office, you can use the following Web sites: www.uspto.com or www.lcweb.loc.gov.

Ironically, a patent doesn't even give one the right to sell a product. It merely excludes anyone else from making, using, or selling your invention. Many business novices who have gone to the trouble to patent a product end up wasting a lot of time and money because a patent is useless if it isn't backed with the right manufacturing, distribution, and advertising programs. As inventor Jeremy

A Proper Copyright Notice

Although a copyright notice is not required by law, you are encouraged to put a copyright notice on every original thing you create. Adding the copyright notice does not obligate you to formally register your copyright, but it does serve to warn others that your work is legally protected and makes it difficult for anyone to claim they have "accidentally stolen" your work. (Those who actually do violate a copyright because they don't understand the law are called "innocent infringers" by the Copyright Office.)

A proper copyright notice includes three things:

1. the word "copyright," its abbreviation "copr.," or the copyright symbol, ©

2. the year of first publication of the work (when it was first shown or sold to the public)

3. the name of the copyright owner. Example: © 2000 by Barbara Brabec. (When the words "All Rights Reserved" are added to the copyright notation, it means that copyright protection has been extended to include all of the Western Hemisphere.)

The copyright notice should be positioned in a place where it can easily be seen. It can be stamped, cast, engraved, painted, printed, wood-burned, or simply written by hand in permanent ink. In the case of fiber crafts, you can attach an inexpensive label with the copyright notice and your business name and logo (or any other information you wish to put on the label).

Gorman states in *Homemade Money,* "Ninety-seven percent of the U.S. patents issued never earn enough money to pay the patenting fee. They just go on a plaque on the wall or in a desk drawer to impress the grandchildren fifty years later."

What a Trademark Protects

Trademarks were established to prevent one company from trading on the good name and reputation of another. The primary function of a trademark is to indicate origin, but in some cases it also serves as a guarantee of quality.

You cannot adopt any trademark that is so similar to another that it is likely to confuse buyers, nor can you trademark generic or descriptive names in the public domain. If, however, you come up with a particular word, name, symbol, or device to identify and distinguish your products from others, you may protect that mark by trademark provided another company is not already using a similar mark. Brand names, trade names, slogans, and phrases may also qualify for trademark protection.

Many individual crafters have successfully registered their own trademarks using a how-to book on the topic, but some would say never to try this without the help of a trademark attorney. It depends on how much you love detail and how well you can follow directions. Any mistake on the application form could cause it to be rejected, and you would lose the application fee in the process. If this is something you're interested in, and you have designed a mark you want to protect, you should first do a trademark search to see if someone else is already using it. Trademark searches can be done using library directories, an online computer service (check with your library), through private trademark search firms, or directly on the Internet through the Patent & Trademark Office's online search service (see checklist and resources). All of these searches together could still be inconclusive, however, because many companies have a stash of trademarks in reserve waiting for just the right product. As I understand it, these "nonpublished" trademarks are in a special file that only an attorney or trademark search service could find for you.

Like copyrights, trademarks have their own symbol, which looks like this: ®. This symbol can only be used once the trademark has been formally registered through the U.S. Patent and Trademark Office. Business owners often use the superscript initials "TM" with a mark to indicate they've claimed a logo or some other mark, but this offers no legal protection. While this does not guarantee trademark protection, it does give notice to the public that you are claiming this name as your trademark. However, after you've used a mark for some time, you do gain a certain amount of common-law protection for that mark. I have, in fact, gained common-law protection for the name of my *Homemade Money* book and successfully defended it against use by another individual in my field because this title has become so closely associated with my name in the home-business community.

Whether you ever formally register a trademark or not will have much to do with your long-range business plans, how you feel about protecting your creativity, and what it would do to your business if someone stole your mark and registered it in his or her own name. Once you've designed a trademark you feel is worth protecting, get additional information from the Patent & Trademark Office and read a book or two on the topic to decide whether this is something you wish to pursue. (See checklist and resources.)

What Copyrights Protect

As a serious student of the copyright law, I've pored through the hard-to-interpret copyright manual, read dozens of related articles and books, and discussed this subject at length with designers, writers, teachers, editors, and publishers. I must emphasize, however, that I am no expert on this topic, and the following information does not constitute legal advice. It is merely offered as a general guide to a very complex legal topic you may wish to research further on your own at some point. In a book of this nature, addressed to hobbyists and beginning crafts-business owners, a discussion of copyrights must be limited to three basic topics:

> **what copyrights do and do not protect**
> **how to register a copyright and protect your legal rights**
> **how to avoid infringing on the rights of other copyright holders**

One of the first things you should do now is send for the free booklets offered by the Copyright Office (see checklist and resources). Various free circulars explain copyright basics, the forms involved in registering a copyright, and how to submit a copyright application and register a copyright. They also discuss what you cannot copyright. Rather than duplicate all the free information you can get from the Copyright Office with a letter or phone call, I will only briefly touch on these topics and focus instead on addressing some of the particular copyright questions crafters have asked me in the past.

Things You Can Copyright

Some people mistakenly believe that copyright protection extends only to printed works, but that is not true. The purpose of the copyright law is to protect any creator from anyone who would use his creative work for his own profit. Under current copyright law, claims are now registered in seven classes, five of which pertain to crafts:

1. *Serials* (Form SE)—periodicals, newspapers, magazines, bulletins, newsletters, annuals, journals, and proceedings of societies.
2. *Text* (Form TX)—books, directories, and other written works, including the how-to instructions for a crafts project. (You

could copyright a letter to your mother if you wanted to—
or your best display ad copy, or any other written words that
represent income potential.)

3. *Visual Arts* (Form VA)—pictorial, graphic, or sculptural
   works, including fine, graphic, and applied art; photographs,
   charts; technical drawings; diagrams; and models. (Also included in this category are "works of artistic craftsmanship
   insofar as their form but not their mechanical or utilitarian
   aspects are concerned.")

4. *Performing Arts* (Form PA)—musical works and accompany-
   ing words, dramatic works, pantomimes, choreographic
   works, motion pictures, and other audiovisual works.

5. Sound Recordings (Form SR)—musical, spoken, or other
   sounds, including any audio- or videotapes you might
   create.

Selling How-To Projects to Magazines

If you want to sell an article, poem, or how-to project to a magazine, you need not
copyright the material first because copyright protection exists from the moment you
create that work. Your primary consideration here is whether you will sell "all rights"
or only "first rights" to the magazine.

The sale of first rights means you are giving a publication permission to print your article, poem, or how-to project once, for a specific sum of money. After publication, you
then have the right to resell that material or profit from it in other ways. Although it is
always desirable to sell only "first rights," some magazines do not offer this choice.

If you sell all rights, you will automatically lose ownership of the copyright to your
material and you can no longer profit from that work. Professional designers often
refuse to work this way because they know they can realize greater profits by publishing their own pattern packets or design leaflets and wholesaling them to shops.

Things You Cannot Copyright

You can't copyright ideas or procedures for doing, making, or building things, but the *expression* of an idea fixed in a tangible medium may be copyrightable—such as a book explaining a new system or technique. Brand names, trade names, slogans, and phrases cannot be copyrighted, either, although they might be entitled to protection under trademark laws.

The design on a craft object can be copyrighted, but only if it can be identified separately from the object itself. Objects themselves (a decorated coffee mug, a box, a tote bag) cannot be copyrighted.

Copyright Registration Tips

First, understand that you do not have to formally copyright anything because copyright protection exists from the moment a work is created, whether you add a copyright notice or not.

So why file at all? The answer is simple: If you don't file the form and pay the fee (currently $20), you'll never be able to take anyone to court for stealing your work. Therefore, in each instance where copyright protection is considered, you need to decide how important your work is to you in terms of dollars and cents, and ask yourself whether you value it enough to pay to protect it. Would you actually be willing to pay court costs to defend your copyright, should someone steal it from you? If you never intend to go to court, there's little use in officially registering a copyright; but since it costs you nothing to add a copyright notice to your work, you are foolish not to do this. (See sidebar, "A Proper Copyright Notice, page 254.")

If you do decide to file a copyright application, contact the Copyright Office and request the appropriate forms. When you file the copyright application form (which is easy to complete), you must include with it two copies of the work. Ordinarily, two actual copies of copyrighted items must be deposited, but certain items are exempt from deposit requirements, including all three-dimensional sculptural works and any works published only as reproduced in or on jewelry, dolls, toys, games, plaques, floor coverings, textile and other fabrics, packaging materials, or any useful article. In these cases, two photographs or drawings of the item are sufficient.

Note that the Copyright Office does not compare deposit copies to determine whether works submitted for registration are similar to any material already copyrighted. It is the sender's responsibility to determine the originality of what's being copyrighted. (See discussion of "original" in the next section, under "Respecting the Copyrights of Others.")

Protecting Your Copyrights

If someone ever copies one of your copyrighted works, and you have registered that work with the Copyright Office, you should defend it as far as you are financially able to do so. If you think you're dealing with an innocent infringer—another crafter, perhaps, who has probably not profited much (if at all) from your work—a strongly worded letter on your business stationery (with a copy to an attorney, if you have one) might do the trick. Simply inform the copyright infringer that you are the legal owner of the work and the only one who has the right to profit from it. Tell the infringer that he or she must immediately cease using your copyrighted work, and ask for a confirmation by return mail.

If you think you have lost some money or incurred other damages, consult with a copyright attorney before contacting the infringer to see how you can best protect your rights and recoup any financial losses you may have suffered. This is particularly important if the infringer appears to be a successful business or corporation. Although you may have no intention of ever going to court on this matter, the copyright infringer won't know that, and one letter from a competent attorney might immediately resolve the matter at very little cost to you.

Mandatory Deposit Requirements

Although you do not have to officially register a copyright claim, it *is* mandatory to deposit two copies of all "published works" for the collections of the Library of Congress within three months after publication. Failure to make the deposit may subject the copyright owner to fines and other monetary liabilities, but it does not affect copyright protection. No special form is required for this mandatory deposit.

Note that the term "published works" pertains not just to the publication of printed matter, but to the public display of any item. Thus you "publish" your originally designed craftwork when you first show it at a craft fair, in a shop, on your Web site, or any other public place.

Respecting the Copyrights of Others

Just as there are several things you must do to protect your "intellectual creations," there are several things you must not do if you wish to avoid legal problems with other copyright holders.

Copyright infringement occurs whenever anyone violates the exclusive rights covered by copyright. If and when a copyright case goes to court, the copyright holder who has been infringed upon must prove that his or her work is the original creation and that the two works are so similar that the alleged infringer must have copied it. This is not always an easy matter, for "original" is a difficult word to define. Even the Copyright Office has trouble here, which is why so many cases that go to court end up setting precedents.

In any copyright case, there will be discussions about "substantial similarity," instances where two people actually have created the same thing simultaneously, loss of profits, or damage to one's business or reputation. If you were found guilty of copyright infringement, at the very least you would probably be ordered to pay to the original creator all profits derived from the sale of the copyrighted work to date. You would also have to agree to refund any orders you might receive for the work in the future. In some copyright cases where the original creator has experienced considerable financial loss, penalties for copyright infringement have been as high as $100,000. As you can see, this is not a matter to take lightly.

This is a complex topic beyond the scope of this book, but any book on copyright law will provide additional information if you should ever need it. What's important here is that you fully understand the importance of being careful to respect the legal rights of others. As a crafts business owner, you could possibly infringe on someone else's designs when you (1) quote someone in an article, periodical, or book you've written; (2) photocopy copyrighted materials; or (3) share information on the Internet. Following is a brief discussion of the first three topics and a longer discussion of the fourth.

1. **Be careful when quoting from a published source.** If you're writing an article or book and wish to quote someone's words from any published source (book, magazine, Internet, and so on), you should always obtain written permission first. Granted, minor quotations from published sources are okay when they fall under the Copyright Office's Fair Use Doctrine, but unless you completely understand this doctrine, you should protect yourself by obtaining permission before you quote anyone in one of your own written works. It is not necessarily the quantity of the quote, but the value of the quoted material to the copyright owner.

   In particular, never *ever* use a published poem in one of your written works. To the poet, this is a "whole work," much the same as a book is a whole work to an author. While the use of one or two lines of a poem, or a paragraph from a book may be considered "fair use," many publishers now require written permission even for this short reproduction of a copyrighted work.

2. **Photocopying can be dangerous.** Teachers often photocopy large sections of a book (sometimes whole books) for distribution to their students, but this is a flagrant violation of the copyright law. Some publishers may grant photocopying of part of a work if it is to be used only once as a teaching aid, but written permission must always be obtained first.

   It is also a violation of the copyright law to photocopy patterns for sale or trade because such use denies the creator the profit from a copy that might have been sold.

3. **Don't share copyrighted information on the Internet.** People everywhere are lifting material from *Reader's Digest* and other copyrighted publications and "sharing" them on the Internet through e-mail messages, bulletin boards, and the like. *This is a very dangerous thing to do.* "But I didn't see a copyright notice," you might say, or "It indicated the author was anonymous." What you must remember is that *everything* gains copyright protection the moment it is created, whether a copyright notice is attached to it or not. Many "anonymous" items on the Internet are actually copyrighted poems and articles put there by someone who not only violated the copyright law but compounded the matter by failing to give credit to the original creator.

   If you were to pick up one of those "anonymous" pieces of information and put it into an article or book of your own, the original copyright owner, upon seeing his or her work in your publication, would have good grounds for a lawsuit. Remember, pleading ignorance of the law is never a good excuse.

   Clearly there is no financial gain to be realized by violating the rights of a copyright holder when it means that any day you might be contacted by a lawyer and threatened with a lawsuit. As stated in my *Crafts Business Answer Book & Resource Guide:*

Changing Things

Many crafters have mistakenly been led to believe that they can copy the work of others if they simply change this or that so their creation doesn't look exactly like the one they have copied. But many copyright court cases have hinged in someone taking "a substantial part" of someone else's design and claiming it as their own. As explained earlier, if your "original creation" bears even the slightest resemblance to the product you've copied—and you are caught selling it in the commercial marketplace—there could be legal problems.

Crafters often combine the parts of two or three patterns in an attempt to come up with their own original patterns, but often this only compounds the possible copyright problems. Let's imagine you're making a doll. You might take the head from one pattern, the arms and legs from another, and the unique facial features from another. You may think you have developed an original creation (and perhaps an original pattern

The best way to avoid copyright infringement problems is to follow the "Golden Rule" proposed by a United States Supreme Court justice: "Take not from others to such an extent and in such a manner that you would be resentful if they so took from you."

Using Commercial Patterns and Designs

Beginning crafters who lack design skills commonly make products for sale using commercial patterns, designs in books, or how-to instructions for projects found in magazines. The problem here is that all of these things are published for the general consumer market and offered for *personal use* only. Because they are all protected by copyright, that means only the copyright holder has the right to profit from their use.

That said, let me ease your mind by saying that the sale of products made from copyrighted patterns, designs, and magazine how-to projects is probably not going to cause any problems *as long*

you might sell), but you haven't. Since the original designer of any of the features you've copied might recognize her work in your "original creation" or published pattern, she could come after you for infringing on "a substantial part" of her design. In this case, all you've done is multiply your possibilities for a legal confrontation with three copyright holders.

"But I can't create my own original designs and patterns!" you moan. Many who have said this in the past were mistaken. With time and practice, most crafters are able to develop products that are original in design, and I believe you can do this, too. Meanwhile, check out Dover Publications' *Pictorial Archive* series of books (see the "Things to Do" checklist and Resources). Here you will find thousands of copyright-free designs and motifs you can use on your craft work or in needlework projects. And don't forget the wealth of design material in museums and old books that have fallen into the public domain. (See sidebar, "What's in the Public Domain?" on page 268)

as sales are limited, and they yield a profit only to you, the crafter. That means no sales through shops of any kind where a sales commission or profit is received by a third party, and absolutely no wholesaling of such products.

It's not that designers and publishers are concerned about your sale of a few craft or needlework items to friends and local buyers; what they are fighting to protect with the legality of copyrights is their right to sell their own designs or finished products in the commercial marketplace. You may find that some patterns, designs, or projects state "no mass production." You are not mass producing if you make a dozen handcrafted items for sale at a craft fair or holiday boutique, but you would definitely be considered a mass-producer if you made dozens, or hundreds, for sale in shops.

Consignment sales fall into a kind of gray area that requires some commonsense judgment on your part. This is neither wholesaling nor selling direct to consumers. One publisher might consider such sales a violation of a copyright while another might not.

Whenever specific guidelines for the use of a pattern, design, or how-to project is not given, the only way to know for sure if you are operating on safe legal grounds is to write to the publisher and get written permission on where you can sell reproductions of the item in question.

Now let's take a closer look at the individual types of patterns, designs, and how-to projects you might consider using once you enter the crafts marketplace.

Craft, Toy, and Garment Patterns

Today, the consumer has access to thousands of sewing patterns plus toy, craft, needlework, and woodworking patterns of every kind and description found in books, magazines, and design or project leaflets. Whether you can use such patterns for commercial use depends largely on who has published the pattern and owns the copyright, and what the copyright holder's policy happens to be for how buyers may use those patterns.

To avoid copyright problems when using patterns of any kind, the first thing you need to do is look for some kind of notice on the pattern packet or publication containing the pattern. In checking some patterns, I found that those sold by *Woman's Day* state specifically that reproductions of the designs may not be sold, bartered, or traded. *Good Housekeeping*, on the other hand, gives permission to use their patterns for "income-producing activities." When in doubt, ask!

Whereas the general rule for selling reproductions made from commercial patterns is "no wholesaling and no sales to shops," items made from the average garment pattern (such as an apron, vest, shirt, or simple dress) purchased in the local fabric store *may* be an exception. My research suggests that selling such items in your local consignment shop or craft mall isn't likely to be much of a problem because the sewing pattern companies aren't on the lookout for copyright violators the way individual craft designers and major corporations are. (And most people who sew end up changing those patterns and using different decorations to such a degree that pattern companies might not recognize those patterns even if they were looking for them. See sidebar, "Changing Things," page 264.)

On the other hand, commercial garment patterns that have been designed by name designers should never be used without permission. In most cases, you would have to obtain a licensing agreement for the commercial use of such patterns.

Be especially careful about selling reproductions of toys and dolls made from commercial patterns or design books. Many are likely to be for popular copyrighted characters being sold in the commercial marketplace. In such cases, the pattern company will have a special licensing arrangement with the toy or doll manufacturer to sell the pattern, and reproductions for sale by individual crafters will be strictly prohibited.

Take a Raggedy Ann doll, for example. The fact that you've purchased a pattern to make such a doll does not give you the right to sell a finished likeness of that doll any more than your purchase of a piece of artwork gives you the right to re-create it for sale in some other form, such as notepaper or calendars. Only the original creator has such rights. You have simply purchased the *physical property* for private use.

How-To Projects in Magazines and Books

Each magazine and book publisher has its own policy about the use of its art, craft, or needlework projects. How those projects may be used depends on who owns the copyright to the published projects. In some instances, craft and needlework designers sell their original designs outright to publishers of books, leaflets, or magazines. Other designers authorize only a one-time use of their projects, which gives them the right to republish or sell their designs to another market or license them to a manufacturer. If guidelines about selling finished products do not appear somewhere in the magazine or on the copyright page of a book, you should always write and get permission to make such items for sale. In your letter, explain how many items you would like to make, and where you plan to sell them, as that could make a big difference in the reply you receive.

In case you missed the special note on the copyright page of this book, you *can* make and sell all of the projects featured in this and any other book in Prima's FOR FUN & PROFIT series.

As a columnist for *Crafts Magazine,* I can also tell you that its readers have the right to use its patterns and projects for money-making purposes, but only to the extent that sales are limited to places where the crafter is the only one who profits from their use. That means selling directly to individuals, with no sales in shops of any kind where a third party would also realize some profit from a sale. Actually, this is a good rule-of-thumb guideline to use if you plan to sell only a few items of any project or pattern published in any magazine, book, or leaflet.

What's in the Public Domain?

For all works created after January 1, 1978, the copyright lasts for the life of the author or creator plus 50 years after his or her death. For works created before 1978, there are different terms, which you can obtain from any book in your library on copyright law.

Once material falls into the public domain, it can never be copyrighted again. As a general rule, anything with a copyright date more than 75 years ago is probably in the public domain, but you can never be sure without doing a thorough search. Some characters in old books—such as Beatrix Potter's *Peter Rabbit*—are now protected under the trademark law as business logos. For more information on this, ask the Copyright Office to send you its circular on "How to Investigate the Copyright Status of a Work."

Early American craft and needlework patterns of all kind are in the public domain because they were created before the copyright law was a reality. Such old patterns may

In summary, products that aren't original in design will sell, but their market is limited, and they will never be able to command the kind of prices that original-design items enjoy. Generally speaking, the more original the product line, the greater one's chances for building a profitable crafts business.

As your business grows, questions about copyrights will arise, and you will have to do a little research to get the answers you need. Your library should have several books on this topic and there is a wealth of information on the Internet. (Just use your search button and type "copyright information.") If you have a technical copyright question, remember that you can always call the Copyright Office and speak to someone who can answer it and send you additional information. Note, however, that regulations prohibit the Copyright Office from giving legal advice or opinions concerning the rights of persons in connection with cases of alleged copyright infringement.

show up in books and magazines that are copyrighted, but the copyright in this case extends only to the book or magazine itself and the way in which a pattern has been presented to readers, along with the way in which the how-to-make instructions have been written. The actual patterns themselves cannot be copyrighted by anyone at this point.

Quilts offer an interesting example. If a contemporary quilt designer takes a traditional quilt pattern and does something unusual with it in terms of material or colors, this new creation would quality for a copyright, with the protection being given to the quilt as a work of art, not to the traditional pattern itself, which is still in the public domain. Thus you could take that same traditional quilt pattern and do something else with it for publication, but you could not publish the contemporary designer's copyrighted version of that same pattern.

10. To Keep Growing, Keep Learning

Everything we do, every action we take, affects our life in one way or another. Reading a book is a simple act, indeed, but trust me when I say that your reading of this particular book *could ultimately change your life.* I know this to be true because thousands of men and women have written to me over the years to tell me how their lives changed after they read one or another of my books and decided to start a crafts business. My life has changed, too, as a result of reading books by other authors.

Many years ago, the purchase of a book titled *You Can Whittle and Carve* unleashed a flood of creativity in me that has yet to cease. That simple book helped me to discover unknown craft talents, which in turn led me to start my first crafts business at home. That experience prepared me for the message I would find a decade later in the book, *On Writing Well* by William Zinsser. This author changed my life by giving me the courage to try my hand at writing professionally. Dozens of books later, I had learned a lot about the art and craft of writing well and making a living in the process.

Now you know why I believe reading should be given top priority in your life. Generally speaking, the more serious you become about anything you're interested in, the more reading you will need to do. This will take time, but the benefits will be enormous. If a crafts business is your current passion, this book contains all you need to know to get started. To keep growing, read some of the wonderful books recommended in the resource section of this book. (If you don't find them in your local library, ask your librarian to obtain them for you through the inter-library loan program.) Join one or more of the organizations recommended. Subscribe to a few periodicals or magazines, and "grow your business" through networking with others who share your interests.

Motivational Tips

As you start your new business or expand a money-making hobby already begun, consider the following suggestions:

- *Start an "Achievement Log."* Day by day, our small achievements may seem insignificant, but viewed in total after several weeks or months, they give us important perspective. Reread your achievement log periodically in the future, especially on days when you feel down in the dumps. Make entries at least once a week, noting such things as new customers or accounts acquired, publicity you've gotten, a new product you've designed, the brochure or catalog you've just completed, positive feedback received from others, new friendships, and financial gains.

- *Live your dream.* The mind is a curious thing—it can be trained to think success is possible or to think that success is only for other people. Most of our fears never come true, so allowing our minds to dwell on what may or may not happen cripples us, preventing us from moving ahead, from having confidence, and from living out our dreams. Instead of "facing fear," focus on the result you want. This may automatically eliminate the fear.

- *Think positively.* As Murphy has proven time and again, what can go wrong will, and usually at the worst possible moment. It matters little whether the thing that has gone wrong was caused by circumstances beyond our control or by a mistake in judgment. What does matter is how we deal with the problem at hand. A positive attitude and the ability to remain flexible at all times are two of the most important ingredients for success in any endeavor.

- *Don't be afraid to fail.* We often learn more from failure than from success. When you make a mistake, chalk it up to experience and consider it a good lesson well learned. The more you learn, the more self-confident you will become.

- *Temper your "dreams of riches" with thoughts of reality.* Remember that "success" can also mean being in control of your own life, making new friends, or discovering a new world of possibilities.

Online Help

Today, one of the best ways to network and learn about business is to get on the Internet. The many online resources included in the "Things to Do Checklist" in the next section will give you a jump-start and lead to many exciting discoveries.

For continuing help and advice from Barbara Brabec, be sure to visit her Web site at www.crafter.com/brabec. There you will find her monthly *Craftsbiz Chat* newsletter, reprints of some of her crafts marketing and business columns, recommended books, and links to hundreds of other art and craft sites on the Web. Reader questions may be e-mailed to barbara@crafter.com for discussion in her newsletter, but questions cannot be answered individually by e-mail.

You can also get Barbara's business advice in her monthly columns in *Crafts Magazine* and *The Crafts Report*.

Until now you may have lacked the courage to get your craft ideas off the ground, but now that you've seen how other people have accomplished their goals, I hope you feel more confident and adventurous and are ready to capitalize on your creativity. By following the good advice in this book, you can stop dreaming about all the things you want to do and start making plans to do them!

I'm not trying to make home-business owners out of everyone who reads this book, but my goal is definitely to give you a shove in that direction if you're teetering on the edge, wanting something more than just a profitable hobby. It's wonderful to have a satisfying hobby, and even better to have one that pays for itself; but the nicest thing of all is a real home business that lets you fully utilize your creative talents and abilities while also adding to the family income.

"The things I want to know are in books," Abraham Lincoln once said. "My best friend is the person who'll get me a book I ain't read." You now hold in your hands a book that has taught you many things you wanted to know. To make it a *life-changing book,* all you have to do is act on the information you've been given.

I wish you a joyful journey and a potful of profits!

"Things to Do" Checklist

INSTRUCTIONS: Read through this entire section, noting the different things you need to do to get your crafts business "up and running." Use the checklist as a plan, checking off each task as it is completed and obtaining any recommended resources. Where indicated, note the date action was taken so you have a reminder about any follow-up action that should be taken.

Business Start-Up Checklist

__Call City Hall or County Clerk

 __to register fictitious business name

 __to see if you need a business license or permit

 __to check on local zoning laws
  (info also available in your library)

 *Follow up:*_____

__Call state capitol

 __Secretary of State: to register your business name;
  ask about a license

 __Dept. of Revenue: to apply for sales tax number

 *Follow up:*_____

__Call your local telephone company about

 __cost of a separate phone line for business

 __cost of an additional personal line for Internet access

 __any special options for home-based businesses

 *Follow up:*_____

__Call your insurance agent(s) to discuss

  __business rider on house insurance
  (or need for separate in-home insurance policy)
  __benefits of an umbrella policy for extra liability insurance
  __using your car for business
  (how this may affect your insurance)

  *Follow up:*_____

__Call several banks or S&Ls in your area to

  __compare cost of a business checking account
  __ get price of a safe-deposit box for valuable business records

  *Follow up:*_____

__Visit office and computer supply stores to check on

  __manual bookkeeping systems, such as the
  Dome Simplified Monthly
  __accounting software
  __standard invoices and other helpful business forms

  *Follow up:*_____

__Call National Association of Enrolled Agents at (800) 424-4339

  __to get a referral to a tax professional in your area
  __to get answers to any tax questions you may have (no charge)

  *Follow up:*_____

__Contact government agencies for information
relative to your business.

  (See "Government Agencies" checklist.)

__Request free brochures from organizations

  (See "Craft and Home Business Organizations.")

__Obtain sample issues or subscribe to selected publications

  (See "Recommended Craft Business Periodicals.")

__Obtain other information of possible help to your business

(See "Other Services and Suppliers.")

__Get acquainted with the business information available to you in your library.

(See list of "Recommended Business Books" and "Helpful Library Directories.")

Government Agencies

__Consumer Product Safety Commission (CPSC), Washington, DC 20207. Toll-free hotline: (800) 638-2772. Information Services: (301) 504-0000. Web site: www.cpsc.gov. (Includes a "Talk to Us" e-mail address where you can get answers to specific questions.) If you make toys or other products for children, garments (especially children's wear), or use any kind of paint, varnish, lacquer, or shellac on your products, obtain the following free booklets:

__*The Consumer Product Safety Act of 1972*
__*The Flammable Fabrics Act*

Date Contacted:_____Information Received:_____

*Follow up:*_____

__Copyright Office, Register of Copyrights, Library of Congress, Washington, DC 20559. To hear recorded messages on the Copyright Office's automated message system (general information, registration procedures, copyright search info, etc.), call (202) 707-3000. You can also get the same information online at www.loc.gov/copyright.

To get free copyright forms, a complete list of all publications available, or to speak personally to someone who will answer your special questions, call (202) 797-9100. In particular, ask for:

__Circular R1, *The Nuts and Bolts of Copyright*
__Circular R2 (a list of publications available)

Date Contacted:_____Information Received:_____

*Follow up:*_____

__Department of Labor. If you should ever hire an employee or independent contractor, contact your local Labor Department, Wage & Hour Division, for guidance on what you must do to be completely legal. (Check your phone book under "U.S. Government.")

Date Contacted:_____Information Received:_____

*Follow up:*_____

__Federal Trade Commission (FTC), 6th Street. & Pennsylvania Avenue., N.W., Washington, DC 20580. Web site: www.ftc.gov. Request any of the following booklets relative to your craft or business:

__*Textile Fiber Products Identification Act*
__*Wool Products Labeling Act of 1939*
__*Care Labeling of Textile Wearing Apparel*
__*The Hand Knitting Yarn Industry* (booklet)
__*Truth-in-Advertising Rules*
__*Thirty-Day Mail Order Rule*

Date Contacted:_____Information Received:_____

Follow up _____

__Internal Revenue Service (IRS). Check the Internet at www .irs.gov to read the following information online or call your local IRS office to get the following booklets and other free tax information:

__*Tax Guide for Small Business—#334*
__*Business Use of Your Home—#587*
__*Tax Information for Direct Sellers*

Date Contacted:_____Information Received:_____

*Follow up*_____

__Patent and Trademark Office (PTO), Washington, DC 20231. Web site: www.uspto.gov

For patent and trademark information 24 hours a day, call (800) 786-9199 (in northern Virgina, call (703) 308-9000) to hear various messages about patents and trademarks or to order the follow-ing booklets:

__*Basic Facts about Patents*

__*Basic Facts about Trademarks*

To search the PTO's online database of all registered trademarks, go to www.uspto.gov/tmdb/index.html.

Date Contacted:_____Information Received:_____

*Follow up:*_____

__Social Security Hotline. (800) 772-1213. By calling this number, you can hear automated messages, order information booklets, or speak directly to someone who can answer specific questions.

Date Contacted:_____Information Received:_____

*Follow up*_____

__U.S. Small Business Administration (SBA). (800) U-ASK-SBA. Call this number to hear a variety of prerecorded messages on starting and financing a business. Weekdays, you can speak per-sonally to an SBA adviser to get answers to specific questions and request such free business publications as:

__*Starting Your Business* —#CO-0028

__*Resource Directory for Small Business Management*—#CO-0042
   (a list of low-cost publications available from the SBA)

The SBA's mission is to help people get into business and stay there. One-on-one counseling, training, and workshops are avail-able through 950 small business development centers across the country. Help is also available from local district offices of the

SBA in the form of free business counseling and training from SCORE volunteers (see below). The SBA office in Washington has a special Women's Business Enterprise section that provides free information on loans, tax deductions, and other financial matters. District offices offer special training programs in management, marketing, and accounting.

A wealth of business information is also available online at www.sba.gov and www.business.gov (the U.S. Business Advisor site). To learn whether there is an SBA office near you, look under "U. S. Government" in your telephone directory, or call the SBA's toll-free number.

Date Contacted:_____Information Received:_____

*Follow up:*_____

__SCORE (Service Corps of Retired Executives). (800) 634-0245. There are more than 12,400 SCORE members who volunteer their time and expertise to small business owners. Many craft businesses have received valuable in-depth counseling and training simply by calling the organization and asking how to connect with a SCORE volunteer in their area.

In addition, the organization offers e-mail counseling via the Internet at www.score.org. You simply enter the specific expertise required and retrieve a list of e-mail counselors who represent the best match by industry and topic. Questions can then be sent by e-mail to the counselor of your choice for response.

Date Contacted:_____Information Received:_____

*Follow up:*_____

Crafts and Home-Business Organizations

In addition to the regular benefits of membership in an organization related to your art or craft (fellowship, networking, educational conferences or workshops, marketing opportunities, etc.), membership may also bring special business services, such as insurance programs, merchant card services, and discounts on supplies and materials. Each of the following organizations will send you membership information on request.

__The American Association of Home-Based Businesses, P.O. Box 10023, Rockville, MD 20849. (800) 447-9710. Web site: www.aahbb.org. This organization has chapters throughout the country. Members have access to merchant card services, discounted business products and services, prepaid legal services, and more.

Date Contacted:_____Information Received:_____

*Follow up:*_____

__American Crafts Council, 72 Spring Street, New York, NY 10012. (800)-724-0859. Web site: www.craftcouncil.org. Membership in this organization will give you access to a property and casualty insurance policy that will cost between $250 and $500 a year, depending on your city, state, and the value of items being insured in your art or crafts studio. The policy includes insurance for a craftsperson's work in the studio, in transit or at a show; a million dollars' coverage for bodily injury and property damage in studio or away; and a million dollars' worth of product liability insurance. This policy is from American Phoenix Corporation; staff members will answer your specific questions when you call (800) 274-6364, ext. 337.

Date Contacted:_____Information Received:_____

*Follow up:*_____

__Arts & Crafts Business Solutions, 2804 Bishop Gate Drive, Raleigh, NC 27613. (800) 873-1192. This company, known in the industry as the Arts Group, offers a bankcard service specifically for and tailored to the needs of the arts and crafts marketplace. Several differently priced packages are available, and complete information is available on request.

Date Contacted:_____Information Received:_____

*Follow up:*_____

__Home Business Institute, Inc., P.O. Box 301, White Plains, NY 10605-0301. (888) DIAL-HBI; Fax: (914) 946-6694. Web site: www.hbiweb.com. Membership benefits include insurance programs (medical insurance and in-home business policy that includes some liability insurance); savings on telephone services, office supplies, and merchant account enrollment; and free advertising services.

Date Contacted:_____Information Received:_____

*Follow up:*_____

__National Craft Association (NCA), 1945 E. Ridge Road, Suite 5178, Rochester, NY 14622-2647. (800) 715-9594. Web site: www.craft assoc.com. Members of NCA have access to a comprehensive package of services, including merchant account services; discounts on business services and products; a prepaid legal program; a check-guarantee merchant program; checks by fax, phone, or e-mail; and insurance programs. Of special interest to this book's readers is the "Crafters Business Insurance" policy (through RLI Insurance Co.) that includes coverage for business property; art/craft merchandise or inventory at home, in transit or at a show; theft away from premises; up to a million dollars in both personal and product liability insurance; loss of business income, and more. Members have the option to select the exact benefits they need. Premiums range from $150 to $300, depending on location, value of average inventory, and the risks associated with one's art or craft.

Date Contacted:_____Information Received:_____

Followup:_____

Recommended Craft Business Periodicals

Membership in an organizations generally includes a subscription to a newsletter or magazine that will be helpful to your business. Here are additional craft periodicals you should sample or subscribe to:

__*The Crafts Report*—*The Business Journal for the Crafts Industry,* Box 1992, Wilmington, DE 19899. (800) 777-7098. On the Internet at www.craftsreport.com. A monthly magazine covering all areas of craft business management and marketing (includes Barbara Brabec's "BusinessWise" column).

__*Craft Supply Magazine*—*The Industry Journal for the Professional Crafter,* Krause Publications, Inc., 700 East State Street, Iowa, WI 54990-0001. (800) 258-0929. Web site: www.krause.com. A monthly magazine that includes crafts business and marketing articles and wholesale supply sources.

__*Home Business Report,* 2949 Ash Street, Abbotsford, B.C., V2S 4G5 Canada. (604) 857-1788; Fax: (604) 854-3087. Canada's premier home-business magazine, relative to both general and craft-related businesses.

__*SAC Newsmonthly,* 414 Avenue B, P.O. Box 159, Bogalusa, LA 70429-0159. (800) TAKE-SAC; Fax: (504) 732-3744. A monthly national show guide that also includes business articles for professional crafters.

__*Sunshine Artist Magazine,* 2600 Temple Drive, Winter Park, FL 32789. (800) 597-2573; Fax: (407) 539-1499. Web site: www.sun shineartist.com. America's premier show and festival guide.

Each monthly issue contains business and marketing articles of interest to both artists and craftspeople.

Other Services and Suppliers

Contact any of the following companies that offer information or services of interest to you.

__American Express. For merchant account information, call the Merchant Establishment Services Department at (800) 445-AMEX.

Date Contacted:_____Information Received:_____

*Follow up:*_____

__Dover Publications, 31 E. 2nd Sreet, Mineola, NY 11501. Your source for thousands of copyright-free designs and motifs you can use in your craftwork or needlecraft projects. Request a free catalog of books in the *Pictorial Archive* series.

Date Contacted:_____Information Received:_____

*Follow up:*_____

__Novus Services, Inc. For merchant account information, call (800) 347-6673.

Date Contacted:_____Information Received:_____

*Follow up:*_____

__Volunteer Lawyers for the Arts(VLA), 1 E. 53rd Street, New York, NY 10022. Legal hotline: (212) 319-2910. If you ever need an attorney, and cannot afford one, contact this nonprofit organization, which has chapters all over the country. In addition to providing legal aid for performing and visual artists and craftspeople (individually or in groups), the VLA also provides a range of educational services, including issuing publications concerning taxes, accounting, and insurance.

Date Contacted:_____Information Received:_____

*Follow up:*_____

__Widby Enterprises USA, 4321 Crestfield Road, Knoxville, TN 37921-3104. (888) 522-2458. Web site: www.widbylabel.com. Standard and custom-designed labels that meet federal labeling requirements.

Date Contacted:_____Information Received:_____

*Follow up:*_____

Recommended Business Books

When you have specific business questions not answered in this beginner's guide, check your library for the following books. Any not on library shelves can be obtained through the library's inter-library loan program.

__*Business and Legal Forms for Crafts* by Tad Crawford (Allworth Press)

__*Business Forms and Contracts (in Plain English) for Crafts People* by Leonard D. DuBoff (Interweave Press)

__*Crafting as a Business* by Wendy Rosen (Chilton)

__*The Crafts Business Answer Book & Resource Guide: Answers to Hundreds of Troublesome Questions about Starting, Marketing & Managing a Homebased Business Efficiently, Legally, & Profitably* by Barbara Brabec (M. Evans & Co.)

__*Creative Cash: How to Profit from Your Special Artistry, Creativity, Hand Skills, and Related Know-How* by Barbara Brabec (Prima Publishing)

__*422 Tax Deductions for Businesses & Self Employed Individuals* by Bernard Kamoroff (Bell Springs Publishing)

__*Homemade Money: How to Select, Start, Manage, Market and Multiply the Profits of a Business at Home* by Barbara Brabec (Betterway Books)

__*How to Register Your Own Trademark with Forms* by Mark Warda, 2nd ed. (Sourcebooks)

__*INC Yourself: How to Profit by Setting Up Your Own Corporation,* by Judith H. McQuown (HarperBusiness)

__*Patent, Copyright & Trademark: A Desk Reference to Intellectual Property Law* by Attorney Stephen Elias (Nolo Press)

__*The Perils of Partners* by Irwin Gray (Smith-Johnson Publisher)

__*Small Time Operator: How to Start Your Own Business, Keep Your Books, Pay Your Taxes & Stay Out of Trouble* by Bernard Kamoroff (Bell Springs Publishing)

__*Trademark: How to Name a Business & Product* by McGrath and Elias (Nolo Press)

Helpful Library Directories

__*Books in Print* and *Guide to Forthcoming Books* (how to find out which books are still in print, and which books will soon be published)

__*Encyclopedia of Associations* (useful in locating an organization dedicated to your art or craft)

__*National Trade and Professional Associations of the U.S.* (more than 7,000 associations listed alphabetically and geographically)

__*The Standard Periodical Directory* (annual guide to U.S. and Canadian periodicals)

__*Thomas Register of American Manufacturers* (helpful when you're looking for raw material suppliers or the owners of brand names and trademarks)

__*Trademark Register of the U.S.* (contains every trademark currently registered with the U.S. Patent & Trademark Office)

Resources

Woodworking Magazines

American Woodworker

P.O. Box 2110
Harland, IA 51593
(800) 666-3111
www.americanwoodworkershow.com

American Woodworker was owned by
Rodale Press for many years, and built
a reputation as an excellent magazine
for woodworkers, providing projects of
varying complexity and techniques, tool
reviews, etc. It was recently bought by
Reader's Digest and it remains to be seen
how they will develop the magazine.

Creative Woodworks and Crafts

243 Newton Sparta Road
Newton, NJ 07860
(973) 383-8080

Primarily scroll sawing, carving, and
intarsia.

The Fine Tool Journal

27 Fickett Road
Pownal, ME 04069
(207) 688-4962
www.wowpages.com/ftj

Devoted strictly to hand tools and
their use.

Fine Woodworking

Taunton Press
63 S. Main St.
Box 5507
Newtown, CT 06470
(800) 688-8286
www.taunton.com

Oldest and best of magazines devoted
strictly to woodworking. Taunton Press
has had a longstanding commitment
to the craft and its advancement. Techniques, tool reviews, projects, etc. In
the past it had a reputation for being
too advanced for beginners, but they
are looking for a broader readership
now. Back issues of this magazine are
one of the best, if not the best, resource
for woodworking techniques.

More Woodturning

950 South Falcon Road
Camano Island, WA 98292
www.fholder.com/woodturning
/woodturn.htm

Tabloid publication devoted strictly
to turning.

Popular Mechanix

P.O. Box 7170
Red Oak, IA 51591
(800) 333-4948
popularmechanics.com

Mostly devoted to other subjects, but always has either woodworking projects, techniques, or tool reviews.

Popular Woodworking

F + W Publications
1507 Dana Ave.
Cincinnati, OH 45207
(515) 280-1721
www.popularwoodworking.com

Project plans, tool reviews, and techniques. Good for the beginner.

Router News

P.O. Box 881, Dept. 88
Brookfield, CT 06804
(203) 775-9290
www.woodwise.com/routernews
/contact.htm

Devoted strictly to using the router.

Shop Notes

August Home Publishing Co.
2200 Grand Ave.
Des Moines, IA 50312
(800) 444-7002
www.augusthome.com/shopntes.htm

Projects and techniques for the hobbyist woodshop.

Weekend Woodcrafts

EGW Publications
1041 Shary Circle
Concord, CA 94518
(800) 777-1164
www.weekendwoodcrafts.com

As the name implies, easier projects that don't take long to make.

Wood Magazine

Better Homes and Gardens
P.O. Box 5049
Boulder, CO 80322
www.woodmagazine.com

Projects, techniques, and tools for the hobbyist woodworker.

Woodshop News

35 Pratt St.
Essex CT 06426
(860) 767-8227
www.woodshopnews.com

Tabloid news format with news for small shops, but plenty of interest to hobbyists such as interviews, tool reviews, and techniques.

Woodsmith

August Home Publishing Co.
2200 Grand Ave.
Des Moines, IA 50312
(800) 444-7002
www.augusthome.com/woodsmth.htm

Well-illustrated project plans for the hobbyist.

Woodwork

Ross Periodicals
P.O. Box 1529
Ross, CA 94957
(415) 382-0580

Project articles, tools and techniques, and interview articles with accomplished woodworkers.

Woodworker West

P.O. Box 452058
Los Angeles, CA 90045
(310) 216-9265
home.woodwest.com

Primarily oriented toward events, activities, and personalities of interest to West Coast woodworkers.

Woodworker's Journal

Rockler Woodworking
4365 Willow Drive
Medina, MN 55340
(612) 478-8201
www.woodworkersjournal.com

Hobbyist magazine with project articles, tool reviews, techniques, etc.

Workbench

August Home Publishing Co.
2200 Grand Ave.
Des Moines, IA 50312
(800) 444-7002
www.augusthome.com/workbnch.htm

From home repair to furniture project articles.

Craft Magazines

The Crafts Report

300 Water St.
Wilmington, DE 19801
www.craftsreport.com

Magazine devoted strictly to the needs of small craft businesses. News, profiles, and articles on all aspects of successfully operating a craft career.

American Craft

American Craft Council
72 Spring St.
New York, NY 10012
(212) 274-0630.
www.craftcouncil.org

Concerns fine arts and crafts. Articles on art criticism and contemporary work.

Suppliers

The major woodworking suppliers sell through mail-order catalogs; many of them are listed here. These suppliers have, until recently, sold only through the mail-order catalogs, but many are now developing Web sites, too.

Ball and Ball Hardware Reproductions

463 West Lincoln Highway
Exton, PA 19341-2594
(610) 363-7330
www.ballandball-us.com

Sells seventeenth- and eighteenth-century reproduction cabinet hardware.

Bridge City Tool Works

1104 N.E. 28th Ave.
Portland, Oregon 97232
(800) 253-3332.
www.bridgecitytools.com

Very high-quality handmade marking tools such as squares, T-bevels, and rulers.

Constantine's

2050 Eastchester Road
Bronx, NY 10461
(800) 223-8087
www.constantines.com

Old established supply house. Along with numerous tools and hardware, you'll find wood, veneers and marquetry supplies, many premade wood parts, project plans, books, and picture frame moldings.

Eagle America

P.O. Box 1099
Chardon, OH 44024
(800) 872-2511
www.eagle-america.com

Router bits, saw blades, and a variety of other tools.

Grizzly Imports

P.O. Box 2069
Bellingham, WA 98227
(800) 541-5537
www.grizzlyindustrial.com

Less expensive import machines as well as a variety of other power tools and hand tools.

Hartville Tool

13163 Market Ave.
Hartville, OH 44632
(800) 345-2396
www.hartvilletool.com

Variety of woodworking tools and supplies.

Highland Hardware

1045 N. Highland Ave. NE
Atlanta, GA 30306
(800) 241-6748
www.highland-hardware.com

Good selection of machine tools, hand tools, and books.

Jesada Tools

310 Mears Blvd.
Oldsmar, FL 34677
(800) 531-5559
www.jesada.com

Router bits and a variety of other small tools.

Lie-Nielsen Toolworks

P.O. Box 9, Route 1
Warren, ME 04864
(800) 327-2520
www.lie-nielsen.com

Makers of high-quality hand planes and other hand tools.

Liechtung Workshops

1108 N. Glenn Road
Casper, WY 82601
(800) 321-6840

Variety of tools and project plans.

Tool Crib

P.O. Box 14930
Grand Forks, ND 58208-4930
(800) 358-3096
www.toolcribofthenorth.com

Wide selection of machines, hand power tools, and other supplies.

Whitechapel Ltd.

P.O. Box 136
Wilson, WY 83014
(800) 468-5534
www.whitechapel-ltd.com

Strictly cabinet hardware. Good selection and prices.

Woodcraft

210 Wood County Industrial Park
Parkersburg, WV 26102
(800) 225-1153
www.woodcraft.com

One of the oldest and best catalogs. Fine selection of carving tools and hand tools, as well as benches, hardware, wood, books, and anything else you could want for your woodworking.

Woodworker's Supply Inc.

1108 North Glenn Road
Casper, WY 82601
(800) 645-9292

A large selection of machines, tools, hardware, books, wood, and finishing supplies.

Woodworking Books

Fortunately, there are a great many books on woodworking out there, from general introductions to highly specialized books on single subjects. Most of the woodworking catalogs sell books, and you'll find advertisements in the woodworking magazines for them. Your local bookstore may have some woodworking books, but generally bookstores carry a minimum of such specialized "limited interest" subjects. But you can directly contact the publishers of such books and ask for their lists, which they will be more than happy to send you. Here is a listing of some of the major woodworking book publishers, along with a few of their offerings:

Betterway Books

F + W Publications
1507 Dana Ave.
Cincinnati, OH 45207
(800) 289-0963

Project and technique books
for the hobbyist.

Make Your Own Jigs and Woodshop Furniture, by Jeff Greef. Designs and instructions for outfitting a shop.

Marvelous Wooden Boxes You Can Make, by Jeff Greef. Wide variety of box-making techniques and designs, along with thorough instructions for making them.

Getting the Very Best from Your Router, by Pat Warner. A router expert tells all.

Sterling Publishing Co.

387 Park Avenue South
New York, NY 10016-8810
(800) 367-9692

Sterling offers a large number of woodworking books from project books to techniques in all major areas of hobbyist interest.

Basics Series, a dozen books on basic techniques for the table saw, router, and band saw, and on sharpening, finishing, cabinetry, and more.

Band Saw Handbook, by Mark Duginske. Very thorough treatment of the band saw.

The Art of Making Elegant Wood Boxes, by Tony Lydgate. Inspiring photos of beautiful boxes, with basic instructions on making them.

Taunton Press

63 South Main St.
P.O. Box 5506
Newtown, CT 06470
(800) 888-8286

Publishers of *Fine Woodworking Magazine,* one of the oldest and most respected wood magazines. They also publish many high-quality books. Here are just a few:

The Workbench Book, by Scott Landis. Definitive treatise on the workbench.

The Router Table Book, by Ernie Conover. Covers a wide range of router table techniques.

The Table Saw Book, by Kelly Mehler. Understanding the capabilities of your table saw.

The Complete Guide to Sharpening, by Leonard Lee.

Understanding Wood, by R. Bruce Hoadley. Excellent book on the basic properties of wood. Essential reading for any beginner.

Books on the Business of Crafts

Crafting As a Business, by Wendy Rosen. The Rosen Group, 3000 Chestnut Ave., Suite 300, Baltimore, MD 21211, (410) 889-2933.

Crafting for Dollars, by Sylvia Landman. Prima Publishing, P.O. Box 1260, Rocklin, CA 95677, (916) 632-4400.

The Crafts Business Answer Book and Resource Guide, by Barbara Brabec, M. Evans.

Creative Cash, by Barbara Brabec. Prima Publishing, P.O. Box 1260, Rocklin, CA 95677, (916) 632-4400.

Craft Shows

The following craft shows are some of the largest and best known in the country. Find smaller, local craft shows by speaking with your local chamber of commerce and galleries and by networking with other craftspeople.

American Craft Council Shows

21 South Eltings Corner Road
Highland, NY 12528
(800) 836-3470
shows@american-craft.org

Largest national crafts organization, operates numerous shows all over the country in Atlanta, Georgia; Baltimore, Maryland; Bellevue, Washington; Charlotte, North Carolina; Chiago, Illinois; San Francisco, California; St. Paul, Minnesota; Tampa Bay, Florida; and West Springfield, Massachusetts. Shows in Spring and Fall, applications due in Fall. High-profile shows where people sell a lot of crafts.

Arrowmont National Show

Arrowmont School of Arts and Crafts
P.O. Box 567
Gatlinburg, TN 37738

Yearly show of wood furniture and crafts from an established school. Applications due in Fall for Spring show.

The Art Furnishings Show

Peck, MacDonald and Assoc.
2899 Agoura Road, Suite 184
Westlake Village, CA 91361
(805) 778-1584
peckmacdon@aol.com

The Art Furnishings Show features furniture, lamps, kitchenware, garden art, wine racks, clocks, and many other items. The show began in April 1999 and takes place in Santa Monica, California, near Los Angeles. Application due date in November. This is a show where you set up a booth to sell your items.

Del Mar Fair Design In Wood Competition

Del Mar Fair Entry Office
P.O. Box 2663
Del Mar, CA 92014

The Del Mar Fair is a large county fair just north of San Diego, California. For years this fair, in conjunction with the San Diego Fine Woodworker's Association, has operated the Design In Wood Exhibition, which takes place in late June and early July every year. It is one of if not the largest woodworking competition and exhibition in the country, with numerous categories in carving, turning, furniture, and more. Applications are due in late April. This is not a show with booths where people sell, but rather a competitive exhibition where you enter your best in hopes of winning a prize. You can sell your piece there, too, if you wish. See the San Diego Fine Woodworker's Association Web site with extensive coverage of the show at www.sdfwa.org.

California Design Exhibition

California Contemporary Craft Association
P.O. Box 2060
Sausalito, CA 94966

An exhibition of contemporary furniture and other objects that takes place once a year in December. Applications due in January. For California residents only. Not a show with booths where you sell, rather an exhibition of pieces, but you can sell what you show if you are accepted.

Paradise City Arts Festival

525 Park Hill Road
Northhampton, MA 01062

Large established shows in Massachusetts where you set up a booth and sell. One in March in Marlborough, and one in June in Northampton. Application deadline for both is November. It features a wide variety of arts and crafts.

Philadelphia Furniture and Furnishings Show

162 North Third St.
Philadelphia, PA 19106
(215) 440-0718

Largest show devoted primarily to furniture in the country. Applications due late Fall for Spring show.

Smithsonian Craft Show

Smithsonian Women's Committee
Smithsonian Institution
A&I Building, Room 1465 MRC 411
Washington, DC 20560-0411
(202) 357-4000

Possibly the most prestigious show in the country, it gets more than ten times as many applicants as it has space for.

Applications due in Fall for Spring show. Wide variety of crafts including furniture and other wood. Here you set up a booth for selling.

Web Sites

There are a variety of different types of Web sites that can be useful to you. Most of the woodworking magazines have sites, and many craft suppliers do as well. You can find craft malls and other online galleries as well. There are numerous organizations that can help you with the business end of your craft. Always look on any Web site you visit for links to other related sites. These will often take you directly to what you are looking for. Here is a sampling of sites of use to you.

Suppliers

The following sites are dedicated Internet businesses that did not start out as mail-order catalog suppliers, like those listed earlier. Many other Web sites also offer supplies for sale as part of their business.

www.wwforum.com
Badger Pond site. Variety of woodworking resources such as articles on wood topics and tool reviews; also features forums, a mall that sells router bits, and links.

www.woodmagazine.com/woodmall
Wood Online
Attn: Marlen
1716 Locust St/GA204
Des Moines, IA 50309-3023

Woodmall site maintained by *Wood* magazine. Online resource for buying tools and other woodworking items.

www.wood-n-crafts.com
Wood-N-Crafts Inc.
P.O. Box 140, Dept. C
Lakeview, MI 48850

Wood-N-Crafts Inc. sells supplies for crafts, but also has a large number of links to other woodworking directories.

Online Craft Malls and Fairs

www.custommade.com
CustomMade.com is a site devoted to connecting fine woodworkers and customers.

www.crafters-market.com

Crafters Market, an online craft mall that charges you a monthly fee and a percentage of sales to post your work on its site.

www.craftsfaironline.com

The Crafts Fair Online has large listings of individual craft artists, shows, supplies, publications, newsgroups, message boards, and more.

Information

www.americancraft.com

The Artist Business Resource site. Extensive information on operating the business end of things.

www.craftassoc.com

The National Craft Association is a national trade group serving the interests of small-scale craftspersons such as yourself. Their address is 1945 E. Ridge Road, Suite 5178, Rochester, NY 14622-2647; (800) 715-9594.

http://artsandcrafts.miningco.com

About.com is a large online encyclopedia of information, this page is about the business of crafts. Contains a large wealth of info and links. Visit this site if you are new to the business of crafts.

http://www.woodworking.miningco.com

About.com's site on woodworking. Extensive resources in all aspects of the craft.

www.1x.com/advisor

Selling Your Art Online by Chris Maher. Newsletter and articles on a wide number of topics around selling online. Good links and resources.

www.crafter.com/brabec

Barbara Brabec's *Craftsbiz Chat*. Online newsletter, articles, and resources for crafts business owners.

www.woodworkersguildofga.org

Woodworker's Guild of Georgia site. Sites from small local organizations like this are a nice change from the glitz and commercialism of larger organizations. Mostly info relevant to the local group; also has extensive links.

www.woodworking.org

Woodworker's Association Web site. A noncommercial site with a great deal of information and links.

www.kiva.net/~rjbrown/w5/wood.html

Woodworking on the World Wide Web site. Numerous resources for woodworkers.

www.woodweb.com

The Woodweb is an information resource for the woodworking industry. Good bookstore listings and links.

www.theoak.com

The Oak Factory site, a woodworking site with very extensive links and other resources.

*www.online96.com/garden/woodwork
.html*

Home and Garden Magazine maintains this extensive list of woodworking links.

Publisher's Sites

www.augusthome.com

August Home Publishing, publisher of *Shop Notes, Workbench,* and *Woodsmith* magazines, provides this online resource for hobbyist woodworkers.

www.blackburnbooks.com

Blackburn Books site, small publisher of quality books. Good book on hand tools and their use.

www.cambiumbooks.com

Cambium Press is the publisher of numerous woodworking books, and this site lists numerous links.

www.taunton.com

Taunton Press, publishers of *Fine Woodworking* magazine. Various resources and links for woodworkers.

Woodworking Glossary

Adze: a handtool used to surface large beams or planks. Like an ax but with the blade turned 90 degrees like a hoe, an adze is used to chop shavings off the surface of the wood leaving a smooth face with characteristic tool marks.

Air dry: lumber is air dry when it has lost all the moisture it can, given the humidity of the surrounding air.

Alaskan chain saw mill: a long chain saw with special attachments used to cut up logs into thick slabs.

Aliphatic glue: common yellow wood-workers glue that is not waterproof but ideal for most interior uses.

Annual rings: the rings you see on a tree stump, also visible on the ends of pieces of lumber.

Arbor: the shaft or spindle on which a blade is placed in a table saw or other woodworking machine.

Architectural work: woodworking pertaining to houses and other buildings, such as doors, windows, and mouldings.

Arkansas stones: sharpening stones made from natural rock found in Arkansas. There are two types, soft and hard. The soft is for medium sharpening and the hard for fine.

Band saw: basic woodworking machine that cuts wood with a thin metal band with cutting teeth.

Bar clamp: clamp with a long bar and adjustable fittings with screw and handle for applying pressure over long distances.

Bearing: fixture on a shaft on which the shaft spins, allowing movement without causing damaging friction. Bearings are also mounted on the tips of some router bits to guide them along the wood edge.

Bench: heavy-duty table on which woodworking procedures are done.

Bench chisel: small cutting tool used mostly for small joinery cuts or bench-work.

Benchwork: joinery and other operations done on a woodworking bench using small hand tools such as chisels, saws, and planes.

Bentwood: a style of furniture based upon bending wood for the major components of the piece.

Bevel: an angle cut on the edge or face of a piece of wood, or on the edge of a cutting tool such as a chisel.

Bimetal blade: special type of band saw blade that has a harder type of steel on the teeth than on the band. The harder teeth stay sharp longer.

Blast gate: valve placed in dust-collecting ducting to shut off or open up the passage of air and collected dust.

Block plane: the smallest commonly used hand plane. Developed for trimming end grain, but useful for numerous small trimmings and adjustments.

Board foot: a basic measure of lumber volume, equal to 1 square foot of wood at 1 inch thick.

Board stretcher: mythical device stolen from Hephaestus by Prometheus, spoken of by many but never actually seen.

Bole: the trunk of a tree from which a saw mill cuts out lumber.

Boring: the act of cutting a hole in wood with a bit.

Bowed: a lumber defect where the wood is curved slightly in a "C" shape.

Brazing: similar to welding, brazing is a means by which metal is "glued" to metal. Router bit tips, or flutes, are brazed onto their metal base.

Built-ins: cabinets that are built into or attached onto the wall as opposed to free standing on the floor.

Burl: odd growth on the surface of a tree, sometimes attaining great size, which yields wood of great beauty.

Burr: small hook on the edge of a scraper that cuts the wood during scraping.

Burring tool: hardened metal shaft used to curl over the edge of a scraper, making a burr.

Butt joint: simple woodworking joint where one piece of wood butts up to another with nothing in between except perhaps glue.

C-clamp: small clamp with a threaded crank for applying pressure over small distances.

Cabinet: literally "small cabin," or little house, any piece of woodworking meant to house objects inside a space behind doors.

Cabriole: curved furniture leg originally meant to imitate the appearance of the leg of a large cat.

Carbide: composite material used on the tips of wood cutting tools such as saw blades and router bits, able to hold a sharp edge 20 times longer than steel.

Carcass: a three-dimensional box; the basic structure of pieces of furniture such as cabinets, desks, or chests of drawers.

Carving: sculptural cutting of wood by hand with special tools.

Carving chisels: specialized chisels that come in myriad shapes and sizes for making any of different shapes of cut required to carve a pattern in wood.

Casting: big chunks of iron that are cast into needed shapes to make the tables and other structures of woodworking machines.

Certification: the process by which a qualified organization ascertains that a given lot of lumber was harvested in a sustainable manner.

Chip breaker: piece of metal on top of a hand plane iron that breaks the chip as the iron peels it off the wood, preventing the chip from tearing the wood ahead of the cut.

Chisel: basic woodworking joinery tool used to cut facets of joints and trim pieces wherever necessary.

Chisel mortising attachment: tool that attaches to a drill press and allows the user to cut square holes in wood to make mortises for mortise and tenon joints.

Chuck: a device that grabs something, such as the chuck in a drill press or hand drill that grabs the drill bit, or on a lathe a chuck grabs the wood and holds it firm as it is turned and cut.

Circular saw: any wood-cutting saw, such as a table saw, hand-held motorized saw, or saw mill with a circular blade that cuts the wood.

Clamp: any of numerous devices that use a mechanical means to apply pressure to a piece or pieces of wood.

Climb cut: with a router, cutting with the rotation of the bit, as if the bit were climbing onto the wood as it cuts.

Collet: small collar that fits around the shank of a router bit and secures it onto the router shaft.

Commercial work: woodworking done for common commercial purposes such as kitchen cabinets.

Common: lesser grades of lumber are called common, such as #1 common or #2 common, decreasing in quality as the number increases.

Contract joinery: cabinetry or millwork done on a contract, such as fitting out a library with all its shelves, built-in cabinets, various moldings, etc.

Countersink: an attachment to a drill bit that bores a hole large enough for the screw head so that the head is recessed below the wood's surface.

Cross cut: any cut on wood that goes across, rather than with, the grain direction.

Crotch: wood that comes from the intersection of branches or crotch on a tree. Such wood is usually highly figured.

Cupped: a wood defect where the board is U-shaped across its width.

Custom: any woodworking done to certain specifications, as opposed to designs that are mass-produced.

Cut: refers to how the board was cut from the tree with respect to the annual rings in the tree, such as flat sawn or vertical.

Cutoff box: jig used on the table saw for the purpose of cutting boards off across the grain, in a cross cut.

Cutter: general term for any sharp piece of metal that cuts wood.

Cutterhead: in a jointer or a planer the cutterhead holds the knives.

Cyanoacrylate glue: instantly bonding glue very handy for zapping down stray splinters that refuse to cooperate otherwise.

Dado: a groove cut in the face, edge, or end of a piece of wood.

Dado cutter: special tool for use on a table or radial arm saw that cuts a wide kerf for making a dado.

Defect: any of numerous features in wood that limit its use, such as knots, splits, or warping.

Design: refers to two aspects of a piece of woodworking—its aesthetics on the one hand and the engineering specifics of its construction on the other.

Dimensions: in woodworking the three critical measurements for each board are its thickness, its width, and its length.

Dovetail joint: basic drawer joint that takes its name from the tapered form of the facets of the joint.

Dovetail jig: specialized tool for use with a router for making dovetail joints.

Dovetail saw: fine-toothed saw used in hand-tool work for cutting dovetail joints by hand.

Dowel: a round peg used to join two pieces of wood together with glue.

Drawknife: a knife blade with a handle on each end used to cut large shavings off the edge of a piece of wood.

Drill press: a machine tool that holds a drill bit vertically and allows the user to push the spinning bit into a piece of wood.

Ducting: tubing made of metal or other materials used to draw air through for the purpose of collecting wood dust and shavings from machines.

Dust collector: a vacuum machine that pulls air through ducting along with wood dust and chips. Filter bags separate the air from the dust and chips.

Dust respirator: a mask worn to reduce the amount of fine wood dust inhaled by someone working with wood.

End grain: the fibers of wood as they appear on the end of a board where they are cut off at 90 degrees.

End split: a split in a board that occurs on the end, usually because the wood was dried too quickly such that moisture was driven out the exposed pores on the end faster than it could leave the middle of the board.

Epoxy: modern adhesive with great strength and gap-filling capabilities, but difficult to clean up and often toxic in formulation.

Exotic: exotic species of wood come from distant lands and have very unusual appearances. Often very desirable, many have been cut close to extinction.

Face mask: dust respirator worn to protect your lungs from dust.

Fair: to trim or shave down a wood surface to smooth it or make its shape more pleasing to the eye.

Fence: any straight attachment that you push wood along to hold it steady as it is cut, as on a table saw or router table.

Fiberboard: a composite sheet made from fine wood fibers held together with glue.

Figure: refers to the aesthetic appearance of wood, such as its color and the patterns created by the annual rings due to the cut of the board.

Firsts and seconds: higher grades of lumber that have a minimum of defects.

Flat grained: boards cut out of the tree such that the annual rings are parallel to the face of the board, or close to it.

Flush: two surfaces that join each other are flush when they are even with each other, with no ridge or step between.

Flush trim: a router bit with straight flutes and a bearing used to cut wood to match a template, causing the edges of the template and wood to which it is attached to become flush.

Flute: the tip on a router bit.

Found lumber: lumber used in the natural state in which it is found, such as driftwood.

Frame and panel: a basic type of construction where a frame is made and a wood panel placed inside the frame.

Fretwork: detailed patterns cut out in thin pieces of wood on a scroll saw for ornamental purposes.

Frog: the frog in your woodshop will not ribbit. It is a metal base in a metal hand-plane that holds the plane iron.

Fullsawn: lumber that has not been planed smooth and still has the saw marks on it from the saw mill.

Glue: any of various liquid formulations used to bond two pieces of wood together.

Gouge: a tool with a U-shaped cutting edge, a gouge is used both in turning and in carving.

Grain: the direction of the wood fibers and pores in a piece of wood, not a reference to their appearance.

Grinder: motorized tool with a grinding wheel used to grind the bevel onto the edge of chisels and plane irons.

Grit: the particles of sand in sandpaper; their size determines the grit size.

Handscrew: wood and metal (or wood and wood) clamp used to apply lighter pressure over a short distance.

Hand tool: traditional tools used by hand to do woodworking operations, or small hand-held motorized tools.

Hardware: metal fixtures such as hinges, locks, and clasps.

Hardwood: wood from deciduous trees, or those that lose their leaves in winter.

Hide glue: traditional wood glue made from the hides of cattle.

Hold-downs: any of numerous devices that do just what the name says.

In-feed table: the table on a jointer ahead of the cutterhead on which the work is fed into the cutterhead.

Jig: any device you make that helps to execute an operation with a tool.

Jig saw: same thing as a scroll saw, sometimes referred to as a saber saw.

Joinery: a broad term referring to the means by which pieces of wood are connected.

Joints: the specific places where pieces of wood are connected together.

Joint: as a verb, to push a piece of wood across a jointer is to joint it.

Jointer: a machine that cuts a straight line on the edge or face of a piece of wood, or a hand plane that does the same.

Keeper: a small piece of wood configured and attached such that it keeps other components in place, as with a table top.

Kerf: the slot or cut made by any saw blade.

Kiln-dried: wood that has been dried in a kiln.

Knotted: wood that has knots, or tree branch bases, visible in the piece.

Lamination: any situation where layers of thin wood are glued together.

Lathe: woodworking machine that spins wood around as it is cut.

Lathe-turned objects: collective term for sculptural objects made on the lathe.

Lacquer: a sprayed on finish developed for ease and speed of use.

Lineal: when you buy wood by the lineal foot you are paying so much for each foot of length you buy, rather than buying by board foot or volume.

Lumber: any wood that has been cut from the tree into long straight pieces.

Lumberyard: a place to tell tall tales of large fish and maybe buy some wood.

Machine tool: a tool that uses an electric motor as opposed to a hand tool.

Mechanical fasteners: screws, nails, and other such metal fasteners that mechanically hold pieces of wood together.

Medium density fiberboard (MDF): a lighter fiberboard useful for simple benchtops or for templates, and in making some cabinets.

Milling: any process of machining wood is generally referred to as milling, from the saw mill to production router setups.

Miter: an angle cut on the end of a board.

Miter gauge: a special sliding fence on a table saw for cutting off pieces on an angle or miter.

Mobile mill: a saw mill that can be brought to a downed tree with a pickup truck.

Mockup: a simple test piece, either of a whole work or part of it, to show proportions, joinery, or some other aspect before actual work is undertaken.

Moisture content: the amount of moisture in a piece of wood.

Molding plane: a hand plane that cuts a specific molding profile onto a piece of wood.

Mortise: a square hole cut in a wood component into which a tenon fits, making a mortise and tenon joint.

Nick: a defect in the cutting edge of a plane iron, planer or jointer knife, or router bit tip. Also refers to the little bump left in the wood by such a defect.

Nominal: woods sold in size gradations are said to be sold in nominal sizes, such as 4, 6, 8, 10, or 12 inches wide. Examples are 2x4 or 1x6.

One-by: the term for lumber that is 1 inch thick. Also written as 1x.

Open time: the length of time that glue remains slippery before it starts to harden.

Orbital: a type of power sander that causes the sandpaper to spin in small circles.

Ornamental turning: a specialized form of lathe work where special tools cut facets in the face of the turning in elaborate patterns.

Outfeed table: the table on the other side of a jointer cutterhead where the work lands after it has been cut.

Overhead costs: the costs of operating a shop that are not specific to a certain project, such a general repairs, rent, and pizza.

Panel: a wide piece of wood, or pieces glued together, placed within a frame in a groove around the inside edge.

Panel raise: a profile cut around the outer edge of a panel where it contacts the frame that gives it depth of appearance.

Pare: to trim or shave down wood, as in benchwork where pieces are hand trimmed to fit.

Paste wax: wax applied over a finish to give a shine.

Plain sawn: flat-grained cut of wood, where the annual rings are parallel to the face of the board.

Plane: as a verb, to use a hand plane or put wood into a planer for the purpose of smoothing the surface or reducing the thickness.

Plane, hand: basic, traditional hand tool for smoothing rough wood and straightening it.

Plane iron: the cutting knife in a hand plane.

Planer: a motorized machine tool for smoothing rough wood or reducing its thickness.

Plywood: a wood-and-glue-manufactured material made by gluing together many thin layers of wood into one large sheet.

Polyurethane glue: a newer type of woodworking glue with excellent sanding and stain-receiving characteristics.

Pores: open tubes or passages in wood that the tree used to transport water and minerals from the roots to the branches. Pores in some woods are large enough to be visible to the eye.

Production: refers to the use of machine tool techniques for the purpose of mass-produced items in a manner that reduces the time per piece required to make components.

Production run: a specific group of items made using production techniques.

Profile: a shape, such as a molding, cut into the edge or face of a piece of wood to make it more attractive.

Push stick: a safety device used at the table saw to push small pieces past the blade without bringing your fingers close to the blade.

Quarter sawn: a cut of lumber where the annual rings of the tree are at 90 degrees to the face of the board, or close to it.

Rabbet: a reduction in the wood, such as a groove in the middle (dado) or on the edge of the board. The English erroneously spell this *rebate*.

Rail: a horizontal component in frame and panel construction.

Radial arm saw: a circular saw mounted on an overhead arm with rollers so that the blade can be pulled across boards.

Random widths and lengths: how hardwoods are sold. The widths and lengths of boards in a lot of hardwood will vary.

Reaction wood: wood that comes from a tree that leaned and had to grow with the wood in tension in order to support itself. This kind of wood is notorious for being unstable and warped.

Reproduction antique: a style of furniture where the maker imitates pieces from the past.

Resaw: splitting boards down their thickness (usually on a band saw) to make thinner boards out of thick ones.

Resorcinol: a two-part glue that is waterproof and tough as nails, if you'll pardon the pun.

Rift grained: a cut of lumber where the annual rings of the tree are at about 45 degrees to the face of the board.

Rip: to cut a board down its length with the grain, as opposed to cutting in across the grain.

Rip saw: a hand saw made specifically to rip lumber, not to cross cut it.

Router: an electric tool that consists of a motor and a base. Cutting tools, called bits, are attached to the end of the motor shaft and spin on it.

Router table: a table top with a router mounted in it such that the bit protrudes up through a hole. Work is pushed across the table top against a fence and into the bit.

Running carving: a type of carving where a repeated pattern is cut into the wood using a series of duplicate cuts.

Rustic: a term used to refer to furniture made from rough lumber such as tree branches and the like.

Saber saw: hand-held electric saw with a reciprocating saber blade sticking out from a small platen. Handy for curved cuts on large pieces.

Salvaged lumber: used lumber taken from old buildings, such as beams, flooring, or even framing lumber, that is given a new lease on life in your woodworking.

Sander: any of numerous power machines that hold sandpaper and use it to scratch wood, none of which should be used very often if you have a sharp smoothing plane and a scraper.

Sanding drum: small rubber cylinder that mounts in your drill press, and has a replaceable sandpaper sleeve on the outside.

Sandpaper: Most insidious invention since the left-handed smoke shifter.

Sapwood: the outer layers of annual rings on most trees have a different color than the inside, and can be of less quality.

Saw: a general term for anything that cuts wood except beavers.

Sawdust: a mysterious substance that magically appears on woodshop floors in direct proportion to the enthusiasm of the inhabitants.

Saw marks: marks left on the surface of wood from any sawing tool.

Saw mill: where trees becomes lumber, either by means of a large circular saw blade, or by means of a large band saw.

Scale drawing: a drawing made to a certain scale, like 1/4 inch to the inch. Such a smaller drawing is easier to make than a full scale drawing.

Schmoo: any sort of liquid or paste used to fill voids in wood, schmoo specifically refers to filling voids that are embarrassing to the craftsperson.

Scorp: a bent drawknife used to hollow out chair seats.

Scraper: thin piece of steel onto which a burr is placed on the edge, for scraping smooth the surface of wood.

Scroll saw: small electric tool with a reciprocating blade that is very thin and so can be used to cut elaborate curved shapes in thin wood.

Section drawing: a drawing of a piece of woodworking looking at a section of it as though you cut it in half. Shows the internal components clearly.

Select: a lumber grade of better quality than common grades.

Setup: an arrangement of tools and/or jigs put together for the purpose of executing a specific operation.

Shaper: motorized woodworking tool with a spindle coming up through a table, onto which you can mount a wide variety of cutters with various shapes.

Sheet goods: plywood, particle board, and other materials that come in 4-by-8-foot sheets, ready for you to cut up on your table saw.

Shellac: traditional finish made from insect cocoons. Seals wood well, looks good, tricky to apply, not resistant to water or alcohol.

Shop: any space where you do your woodworking.

Skew: a tool for cutting wood on the lathe that has a straight cutting edge set at an angle to the length of the tool.

Smoothing plane: smaller hand plane used specifically for putting a finished surface on wood.

Softwood: wood that comes from coniferous trees, which generally are trees with needles rather than leaves and which stay green in winter.

Sole: a plane sole is the flat bottom of the tool that rides on the wood as it is being cut.

Solvent: liquid substance that dissolves other substances, such as water dissolves many glues, or alcohol dissolves shellac, or lacquer thinner dissolves lacquer.

Spline: a thin piece of wood placed in grooves in two pieces of mating wood, which acts as a tenon to join the pieces together.

Spokeshave: a small handtool with a knife mounted in it which you draw across the surface, pulling a thin shaving. Good for shaping curves.

Stickers: skinny sticks of wood that you put between green wood or any wood stored outside so that air can circulate around it and the wood can dry.

Stile: the vertical components in a frame and panel construction.

Stile and rail profile: the molded pattern used along the inside edge of a frame.

Stop block: a block clamped onto a fence to fix the location of the ends of pieces of wood with respect to the cutting tool.

Stopped cut: a cut that is stopped in the middle of a board rather than cut all the way through it.

Strap clamp: a clamp made with a strong cloth strap and a small winch that pulls the strap tightly around an object.

Surface: as a verb, to surface means to put a smooth surface on a piece of lumber where there was a rough surface before, by planing (hand or machine) or sanding.

Surfaced lumber: lumber that has had the saw marks removed from it so that it is smooth.

Sustainable harvesting: forestry and lumber cutting practices that leave behind a forest that can rejuvenate itself to a certain extent.

Swirl marks: why you shouldn't use sanders. Orbital sanders can leave swirl scratches in the wood surface that are harder to remove than learning to use a scraper.

Table, machine: cast iron table top on a woodworking machine upon which you place work as it is machined.

Table saw: basic woodworking machine with a circular saw blade coming up through the top of the machine.

Taper: sloping shape on a piece of wood like a wedge.

Tear-out: surface defect on the wood caused by cutting tools pulling the fibers up from below the surface.

Template: a pattern made of plywood or other material used to trace or flush trim a curved shape onto a piece of wood.

Tenon: a finger of wood that fits into a mortise, forming a mortise and tenon joint for connecting two pieces of wood.

Trestle: a type of table leg and rail construction that uses mortise and tenon joinery.

Tung oil: a high quality oil used in some oil-based finishes.

Turner: famous media mogul. Or any person who uses a lathe to make turnings.

Turning: that which is made by turning on a lathe.

Two-by: name by which 2-inch thick lumber is called. Also written 2x.

Unsurfaced lumber: rough lumber as it comes from the saw mill with the sawmarks still on the wood.

Urea formaldeyde: common woodworking glue with a long open time and high water resistance.

Veneer: a very thin piece of beautiful wood glued onto a substrate to make an attractive surface.

Vertical grained: quarter-sawn lumber, or wood that has the annual rings at 90 degrees to the face of the board.

Warp: a defect in wood where the board is twisted down its length.

Whittle: using a knife to carve a piece of wood. By extension, you could refer to all woodworking as whittling.

Wood movement: the effect of changes in moisture content in wood. As the wood takes on moisture it expands, and contracts as it loses it.

Index

About the Author

JEFF GREEF has written three books and over a hundred magazine articles for hobbyist woodworkers on subjects ranging from furniture projects to tool reviews and interviews with other woodworkers. He makes his home in Santa Cruz, California.

About the Series Editor

BARBARA BRABEC is one of the world's leading experts on how to turn an art or crafts hobby into a profitable home-based business. She regularly communicates with thousands of creative people through her Web site and monthly columns in *Crafts Magazine* and *The Crafts Report*.

To Order Books

Please send me the following items:

Quantity	Title	U.S. Price	Total
_____	Decorative Painting For Fun & Profit	$ 19.99	$ _____
_____	Holiday Decorations For Fun & Profit	$ 19.99	$ _____
_____	Woodworking For Fun & Profit	$ 19.99	$ _____
_____	Knitting For Fun & Profit	$ 19.99	$ _____
_____	Quilting For Fun & Profit	$ 19.99	$ _____
_____	Soapmaking For Fun & Profit	$ 19.99	$ _____
_____	_____	$ _____	$ _____
_____	_____	$ _____	$ _____

Subtotal	$ _____
Deduct 10% when ordering 3–5 books	$ _____
7.25% Sales Tax (CA only)	$ _____
8.25% Sales Tax (TN only)	$ _____
5% Sales Tax (MD and IN only)	$ _____
7% G.S.T. Tax (Canada only)	$ _____
Shipping and Handling*	$ _____
Total Order	$ _____

*Shipping and Handling depend on Subtotal.

Subtotal	Shipping/Handling
$0.00–$29.99	$4.00
$30.00–$49.99	$6.00
$50.00–$99.99	$10.00
$100.00–$199.99	$13.50
$200.00+	Call for Quote

**Foreign and all Priority Request orders:
Call Customer Service
for price quote at 916-632-4400**

This chart represents the total retail price of books only
(before applicable discounts are taken).

By Telephone: With American Express, MC, or Visa,
call 800-632-8676 or 916-632-4400. Mon–Fri, 8:30–4:30.
www.primapublishing.com
By E-mail: sales@primapub.com
By Mail: Just fill out the information below and send with your remittance to:
Prima Publishing ▪ P.O. Box 1260BK ▪ Rocklin, CA 95677

Name _____

Address _____

City_____ State _____ ZIP _____

MC/Visa/American Express# _____ Exp._____

Check/money order enclosed for $ _____ Payable to Prima Publishing

Daytime telephone _____

Signature _____